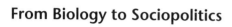

From Biology to Sociopolitics

From Biology
to Sociopolitics

Conceptual Continuity in

Complex Systems

Heinz Herrmann

Yale University Press

New Haven & London

Printed in the United States of America.

Library of Congress Cataloging-in-Publication Data

Herrmann, Heinz, 1911–
From biology to sociopolitics : conceptual continuity in complex systems / Heinz Herrmann.
    p.   cm.
Includes bibliographical references and index.
ISBN 0-300-07253-8 (alk. paper)
  1. Chaotic behavior in systems. 2. Conceptual structures (Information theory) 3. Complexity (Philosophy) 4. Simplicity (Philosophy) I. Title.
Q172.5.C45H47 1998
003—dc21   97-51555
                                                  CIP

A catalogue record for this book is available from the British Library.

The paper in this book meets the guidelines for permanence and durability of the Committee on Production Guidelines for Book Longevity of the Council on Library Resources.

10 9 8 7 6 5 4 3 2 1

This book is dedicated to the staff of the Study Circles Resource Center, who are transforming the idea of conceptual continuity into understanding and cooperation between sociopolitical opposites

Contents

Preface

The need to deal with complexity has become one of the salient characteristics of the postmodern era—the second half of the twentieth century. During this time the extraordinary advances in cell and molecular biology have opened the black box of the living system, revealing an unparalleled array of diverse components and their specific interactions on several levels of a hierarchical and highly complex organization. This was accomplished in a step-by-step analysis, over several decades, and without direct reference to a unifying, general theory.

Even though incomplete, our present knowledge of living systems is sufficiently advanced to ask in which sense we have achieved an understanding of the complexity of living matter. If a particular mode of thinking leads to an understanding of the complexity of the living state, does it apply to other system types as well? In seeking an answer to these questions, this book attempts to create a perspective in which the inanimate world, life, and human existence are envisioned as systems of increasing complexity representing physical, biological and sociopolitical realities. These systems are tentatively thought to be related by a common form of understanding, here called conceptual continuity. This conceptual continuity is established when two entities can be shown to form an intermediary complex or to share a common element. To give this idea concrete meaning right away, we can think of the mathematical relationship between temperature and fluid expansion in a thermometer. Conceptual continuity is established through the introduction of molecular motility. With increasing temperature, the movement of the molecules of the thermometer fluid increases in proportion to the increase in temperature. At the same time the increase of molecular movement causes the fluid to expand. Molecular movement is therefore the element shared by temperature and fluid expansion that establishes conceptual continuity between these two variables. Similarly, the DNA-messenger-RNA complex establishes conceptual continuity for one decisive step in the process of gene expression. The actin-myosin complex is the common link that establishes conceptual continuity between muscle structure and function. Likewise, in the sociopolitical realm, careful negotiations under the leadership of two unique personalities—Nelson Mandela and F. W. de Klerk—led to the abandonment of Apartheid with a beginning of conceptual continuity between opposed races.

Up to the middle of this century, systematic studies of complexity were avoided since the simplifying generalizations of physics had generated an apparently adequate account of the diverse phenomena of the universe. Since that time, various types of theories of complexity have been developed, such as systems theory or chaos theory. Obviously, without general theories one cannot conceive of advancements in science. The relevance, however, of general theories for a complete understanding of systems of different complexity may not be the same. Therefore, conceptual continuity is examined in the following chapters as a complementary, alternative mode of understanding that applies to systems of widely differing character and complexity. This survey will cover physical systems, certain biological systems, sociopolitical systems, and certain historical aspects that indicate the range and the meaning of conceptual continuity. The possibility that a recognition of conceptual continuity may contribute—if only infinitesimally—to an understanding of our staggering sociopolitical problems, is one of the primary reasons for writing this book.

Acknowledgments

The writing of this cross-disciplinary book was made possible only through the support of my colleagues at the University of Connecticut. I am indebted to Professors Hugh Clark, Julius Elias, Anne Hiskes, Hallie Krider, Petter Juel Larsen, and Robert Vinopal for their encouragement, discussions, and suggestions.

I am grateful to Professor J. A. Janik (Department of Physics, University of Krakow), who offered critical comments after reading the manuscript during a visit to the United States, and to Robert Stallman for essential modifications of the Introduction.

I want to convey special thanks to Dr. Michael Lefor for extensive and incisive editorial scrutiny of the manuscript, to Elizabeth Jean for the highly competent word processing of the material, and to Philip Marcus for making his computer facility available.

My introduction by Martha McCoy, Executive Director of the Study Circles Resource Center, to the academic and civic aspects of deliberative democracy was invaluable.

I am glad to acknowledge with profound appreciation the enthusiastic reception of the manuscript by Professor Everett Ladd, Director of the Roper Center for Public Opinion Research at the University of Connecticut, which launched the manuscript into the sphere of the publisher.

I regard the effective collaboration with Jean Thomson Black, the Science Editor, and the staff of Yale University Press, as a great privilege.

Introduction

The gap between the perception of the physical universe and the experience of human existence has never been so great as in recent times. Beginning in ancient Greece, the human mind developed a basic need to encompass the diversity of nature by substituting simplifying abstractions (Chapter 1). Following the Copernican Revolution and the establishment of Galilean-Newtonian mechanics, scientific thought has tended toward subsuming the diversity of phenomena under a single "Final Theory" (Weinberg 1993). These advances have engendered the attitude that all forms of understanding, even those pertaining to life and human existence, should follow some type of simplifying abstraction and that those introduced by physics are the only models of understanding worth consideration.

Contrary to this view, the analysis of the basic life processes—reproduction, genetic transmission, motility, responsiveness, synthesis of complex molecules, evolution, embryonic development—revealed that the maintenance of living matter involves the complexity of highly specific interactions between a great number of particular cell-constituents, cells, and tissues. Similarly, a generalized and simplified perception of individual or collective human existence has become insufficient, considering the wide range of events in many fields of human endeavor in all corners of the world and the number and gravity of sociopolitical problems endlessly exposed by the media.

Are we thus facing two separate views of reality, classified as "simple" and "complex"? Does a simple-complex dichotomy separate physics from large areas of biology and sociopolitics by making them noncomparable in any substantive, mutually interpretable sense?[1] Is there a form of understanding that is common to both types of reality?

In considering these questions this introduction first briefly restates the elementary characteristics of simple and complex forms of reality as a basis for later discussions. This leads to an examination of the nature and the use of conceptual continuity for the understanding of different forms of reality and to speculations about the role of different models of reality in the management of human society. To facilitate comparison, the main characteristics of the discussed systems are summarized in Table I.1 as elaborated in more detail in the subsequent chapters.

Turning first to the physical models, we can identify single unit systems that include an infinite number of randomly distributed elements such as the molecules in volumes of gas or liquid or, on a different scale, the stars in the universe. These are the systems of statistical mechanics and astrophysics, respectively. Systems that include a limited number of ordered components belong to the realm of mechanics. Changes in both types of systems are fully defined by mathematical equations (Table I.1, IA–IC; Chapters 2 and 3). For example, all forms of macroscopic motions at speeds below that of light are subsumed under a few concepts of mechanics (mass, force, acceleration). These systems are isolated in thought from the rest of the concrete universe and are conventionally regarded as *closed systems.* By definition, any characteristics that are not included in such formal representations can be disregarded. One can speak here of *fully defined* or *ideal systems* that represent, in this sense, reality in its simplest form. For example, the Newtonian equations describe the state of the planetary system in terms of mass, motion, and gravitational force and ignore the possibility of complicating factors, such as a reversal of the expansion of the universe, an atomic explosion of the sun, or the impact of meteors on Jupiter.

In contrast to the ideal simplifying, generalized representations of the planetary systems, explorations that differentiate the planets deal with their specific properties. Our continuing scrutiny of the earth, of its atmosphere, oceans, and landmasses, with their deserts, rainforests, and icecaps, presents the earth as an example of nonideal systems. These are systems comprising varying and often large numbers of heterogeneous components interacting with one another and their environment with differing degrees of specificity. Not all factors that determine the state of the system are identified. Hence, we can speak of *open, only partially defined,* or *nonideal systems* with a higher degree of complexity than ideal systems (Table I.1, IIA–IIB; Chapter 4).

A juxtaposition of ideal and nonideal systems acquires particular relevance in understanding living systems and is given special attention here (Table I.1, IIC; Chapter 4). For several centuries it was thought that the peculiarities of life, such as reproduction, growth, motility, and responsiveness to external stimuli, could not be explained by the concepts of physics. However, the recognition of the conservation of energy in living systems, the possibility of defining metabolic processes in terms of thermodynamics, the mathematical expression of biological growth and form, the physico-chemical treatment of the permeability of biological membranes, the mechanistic qualities of muscular contraction, and the change of electrical charges as a basis of nerve conduction supported the validity of an analysis of life processes that was in accordance with the simplifying, general principles of physics. As a result, large areas of biology were treated as a subdivision of physics up to the halfway mark of the present century. This subordination lost its relevance during the course of subsequent developments in the analysis of living systems.

Interposed between the subatomic-atomic and the planetary extremes of comprehension, the dimension of the living state coincides with the range of our direct sensory perception. Together with light and electron microscopy, a host of easily handled methods of experimental analysis has revealed a large number of diverse, specific constituents of living systems, and the reconstruction of the living state from its components has dramatically advanced during the last half of this century. With the identification of the macromolecular components of living organisms, it became apparent that their highly specific interactions give rise to the organization of subcellular elements, of these elements into cells, and of cells into tissues and organs. Furthermore, it became known how cells respond to signaling between organs by hormones and neurotransmitters; how the transmission of hereditary characteristics and their expression during embryonic development is controlled; how light is used by plants in producing living matter; and how nutrients are converted into metabolic energy for the maintenance and activation of functions on all levels of organization. All this basic information promoted significant advances in agriculture, medicine, and biotechnology. Obviously, much of this is now being taken for granted. The DNA double helix has become a household word and genetic engineering is a topic for talk shows. Thus, the headlong rush toward recognition of the full details of the living state was not conducive to gaining an overview by asking questions about a deeper meaning of this rather unique development.

To reiterate, in the closed systems of physics the smallest possible number of general variables are related by mathematical equations in accordance with theories that can be experimentally verified. The verified theory is then regarded as the most adequate form of understanding of physical systems. In contrast, the analysis of living systems identifies the largest possible number of specific components and depends on a step-by-step recognition of their highly specific interactions. Neither a general mathematical formalization of these interactions nor a covering theory was needed for the advance of this analysis.[2] As suggested in the Preface, an alternative form of understanding that applies to any level of complexity, including physical, living, and sociopolitical systems, is the establishment of *conceptual continuity.*

Conceptual continuity between two entities is obtained when an element is recognized that is shared by hitherto unrelated, empirically correlated, or formally-mathematically (algorithmically) related system components (for concrete examples, see Preface; Chapters 3 and 4). The advantage of the idea of conceptual continuity is its independence of the complexity of the systems. Thus, conceptual continuity assumes three forms corresponding to the three classes of systems listed in Table I.1. The three classes refer to: high relevance of simplifying abstraction, intermediate relevance of simplifying abstraction, and low relevance of simplifying abstraction. The first class is characteristic of ideal systems; the second and third classses are characteristic of nonideal systems. In the ideal, physical systems conceptual continuity is established by a single, general, linking unit. For example, molecular motion is the common element of mechanical work and heat; photons exist as an intrinsic part of both a beam of light and of the atoms of the light-absorbing material (Chapter 3).[3]

In nonideal geological or meteorological systems the specific conditions producing the effects of general mechanical (for example, continental drift) or thermodynamic forces (hurricanes) become the links establishing conceptual continuity. In biological systems highly specific molecular complexes link biological phenomena, such as gene expression or muscle contraction (see Chapter 4). In the field of sociopolitics the recognition of even small areas of common interest between individual and society or opposing social groups point to the crucial linking units in the establishment of conceptual continuity (Chapters 5 and 7).

Democratic government, in particular, attempts to establish conceptual continuity among the social components of its organization by developing areas of common interest among generations, social classes, and races and by preventing ideological polarization between political parties (Chapter 5). In contrast, conceptual discontinuity in sociopolitical systems is represented by adversarial ideologies of political parties, nations, races, or social classes. Here the individual will pay for a reduction in the demands of a democratic society with a concomitant decline in the freedom of expression and related privileges and with a diminished opportunity for the realization of conceptual continuity in the radically ideological forms of totalitarianism, nationalism, communism, or fundamentalism (Chapter 6). The foregoing examples show that the establishment of conceptual continuity requires the identification of the element and the process that joins two hitherto separate entities. This can be regarded as defined conceptual continuity. In contrast, in some phases of the philosophical thought of ancient Greece and of the Enlightenment hypothetical relationships between entities were proposed in which the linking process remained unidentified; therefore, conceptual continuity was not achieved. In this case one is dealing with conjectural or hypothetical correlations.

The perception of the simplifying generalizations and complex specificities and of the role of conceptual continuity has changed during consecutive periods of Western society, giving the simple/complex relationship a time dimension. In science the relationships between the parts of a system are explored by recording the effects of experimentally subjecting the systems to a variety of well-defined conditions. Such systematic experimentation is not possible with human societies. But by simulating experimentation, history provides the changing conditions and records their effects on human society.

Briefly summarized, concrete forms of high and low levels of sociopolitical complexity are represented by the quasi-democracy of Athens and the totalitarianism of Sparta, respectively. Achieving a compromise between two opposing social classes, Solon created one of the earliest examples of sociopolitical conceptual continuity by drafting the Athenian constitution. In Spartan history, Lycurgus terminated a poorly understood period of individual freedom by his introduction of a totalitarian form of governance and a concomitant reduction of the extent of conceptual continuity (Chapter 1).

Abstract simplifications first appear among the surviving fragments of the pre-Socratic thinkers of ancient Greece in the form of speculations about the nature of the universe. Ostensibly, a concern with nature was also part of the ancient Babylonian, Egyptian, Hindu, and Chinese civilizations, but only the ancient Greeks attempted to develop a rational type of natural philosophy (Blanpied 1969, p. 3).[4] In one view, the objects of nature were thought to be varying mixtures of four elements: earth, water, fire, and air. In an alternative interpretation, the universe was conceived to be founded on a few basic, geometrically perfect structures, suggesting merely a hypothetical correlation.

Further development of philosophical thinking by Plato and Aristotle led to attempts to deal with human existence. This gave rise to the basic dichotomy of conceiving human society either as a simplified, abstract utopia—as in Plato's *Republic* and Aristotle's *Nicomachean Ethics*—or as a complex system of specific prescriptions for the day-to-day conduct of human affairs, as given in Plato's *Laws* and Aristotle's *Politics* (Chapter 1).

During the Middle Ages the burdens of a transient, earthly reality were made tolerable by the promise of a Kingdom of God and the possible return to Paradise in an eternal hereafter. The subsequent period, from the Copernican revolution to the publication of Newton's *Principia,* was the beginning of modernity, as mechanistic physics initiated a scientific simplification of reality with the direct or indirect transmission of force as the tenuous basis of conceptual continuity. Mechanistic physics became a model not only for other sciences but—positively or negatively—influenced systematic thought about human society as well. In that context, visionary or speculative simplifying generalizations of sociopolitical systems, later called utopias, attained a prominent role in philosophical thinking (Chapter 2). Inspired by the advances of physics, Hobbes and the French *Philosophes* based utopias on general mechanistic principles as simplifying forms of human existence, with Locke questioning the validity of this approach. Somewhat later, Rousseau opposed mechanistic science as a basis for human existence and created a powerful utopian simplification by invoking an egalitarian form of governance and economy with conceptual continuity remaining problematical. However, even Rousseau had to yield to the simple/complex dichotomy. When he was asked for guidance in planning governance in concrete situations and proposed constitutions for Corsica and Poland, Rousseau relinquished his utopian simplifications and introduced complex forms of governance. It should also be emphasized that at about the same time Montesquieu and Burke developed an empirical, pragmatic approach to governance by considering specific conditions as the basis for conceptual continuity of complex constitutionalism (Chapter 2), disregarding the need for utopian simplifications and abstract generalizations in the conduct of human affairs.

Evidently, several main categories of the simple/complex continuum existed by the end of the Enlightenment. Subsequently, these categories acquired greater distinctness in the nineteenth and twentieth centuries as separate simple or complex forms of physical, biological, and sociopolitical reality. The simplifications of Newtonian physical reality were thus greatly expanded (Chapter 3). It was during this period that the interconvertibility of the different forms of energy was demonstrated. Also, the apparent disparities of physical phenomena, such as light and matter, were replaced by conceptual continuity, with photons as intrinsic and exchangeable components of both. Spectral analysis of light ushered in the understanding of atomic structure, which in turn provided a unifying explanation of a wide range of physical and chemical processes. Later investigations led to the clarification of a theory of radioactivity and its relationship to the structure of the atomic nucleus and eventually to an understanding of the relationship of atomic energy to many of the properties of the universe, as mentioned in Chapter 3. This resulted in a further unification of physical reality. In its most recent phase, deviating from its course in preceding periods, science began to deal with complex, nonideal forms of reality, the heterogeneities of the universe, geology, meteorology, and biology (Chapter 4).

During the past two centuries two divergent sociopolitical systems developed. Highly complex systems of governance—constitutionalism and democracy—became models of sociopolitical reality with many possibilities for conceptual continuity (Chapter 5). In contrast—avoiding an increase in sociopolitical complexity—Nationalism and Communism diminished the extent of conceptual continuity through the simplifying ideologies of totalitarian systems. The coexistentiality and tolerance of the social utopias of the Enlightenment were replaced by an intolerant noncoexistentiality of totalitarian rule (Chapter 6).

In the past, discussions of the relevance of science for the interpretation of reality, including human existence, were based on the simplifications of physics. However, the close relationship of alternative interpretations of living matter (in terms of either simplifying generality or complex specificity and its extensive network of conceptual continuity) suggests that the understanding of living, rather than inanimate systems, would be of greater use in reconciling the simplifying and complex qualities of reality. With its emphasis on specificity, biology becomes a focal point in the discussion of the simple/complex relationship, a pivotal area in bridging the gap between the abstract realities of mechanistic physics and the concrete realities of human society. It legitimizes biological systems as models for establishing conceptual continuity as a way of understanding complexity. Is the complexity of democracy, engendering a resilience comparable to that of life, a decisive advantage, or are at least some forms of democratic complexity ultimately self-destructive? Are the simplifying ideologies of totalitarianism offering possibilities of escape from the burdens of democratic complexity? The search for conceptual continuity in the creation of social patterns that will accommodate complexity as an intrinsic form of reality is a major challenge for our time.[5]

Table I.1 Tentative Classification of Systems of Increasing Complexity

General nature of system	Specific type of system	Theories	Order	Thermodynamic dependence
			I. Ideal systems—fully defined (simple) High relevance of simplifying abstractions Conceptual continuity established by general elements	
IA. Infinite number of randomly distributed parts	Solutions, gases	Statistical mechanics; system fully defined	No order; unspecific interaction between parts through thermal collision	Thermodynamic equilibrium
IB. Limited number of parts with large-scale spatial relations and relatively high stability	Planetary system	Newtonian mechanics; system fully defined	Unspecific interaction of parts by gravitational force; infinite number of systems possible; no restraints	No thermodynamic dependence
IC. Limited number of parts with micro-scale discrete spatial relations	Atomic and molecular structure, electron orbitals	Quantum mechanics; wave mechanics; periodic law; system fully defined	Order based on mutual restraints between electron orbitals; specific atomic interactions; only a limited number of possible structures	No thermodynamic dependence

(continued)

Table I.1 Continued

General nature of system	Specific type of system	Theories	Order	Thermodynamic dependence
		II. Nonideal systems—not fully defined (complex) Intermediate relevance of simplifying abstractions Conceptual continuity established by general elements modified by specific conditions		
IIA. Nonequilibrium physical systems	Convective solutions	Irreversible nonequilibrium thermodynamics; Chaos theories	Dynamic nonrandomness	Continuous energy requirement
IIB. Limited number of closely integrated components	Earth, other individual planets, and sun	No overall theory; theories only about components (continental drift); system not fully defined	Highly structured surface, geological stratification; oceanic, atmospheric currents; nonspecific interactions between a large number of different components	Continuous uptake of solar radiation and release of energy by heat dissipation; generation of steady-state heat produced by radioactive decay in earth's interior

		Low relevance of simplifying abstractions	Conceptual continuity established by specific elements	
IIC. Very large number of highly organized parts; multiple levels of organization	Cells as the smallest units of living systems	No overall theory; molecular definition of isolated systems; partial generalizations (e.g., genetics)	Highly specific interaction of parts with self-assembly into labile functional cell components; informational structures; hierarchial organization	Continuous uptake of energy, in part for synthesis of macromolecules steady state with high negative entrophy; dissipation of energy as metabolic waste and heat
IID. Large numbers of human individuals forming associations with diverse types of organization	Human societies	Speculative utopian simplifications. Nonutopian pragmatic complex representations	Wide range from more highly ordered totalitarian to less highly ordered democratic societies	Variable but mostly high rates of energy conversion

Chapter 1 The Dawn of the Simple-Complex Dichotomy and of Conceptual Continuity

It may seem incongruous to open a discussion of complexity as a characteristic of the postmodern era with a survey of the sociopolitical organization of ancient Greece and of the main achievements of its great thinkers. Actually, the history of ancient Greece describes a remarkable set of sociopolitical experiments that produced different forms of concrete sociopolitical complexity. During the late phase of Greek antiquity Plato and Aristotle developed abstract schemes for governance as part of their philosophical thinking. That both the sociopolitical reality of that time as well as its philosophical representation are relevant for the understanding of modern complexity has been emphasized by Sir Ernest Barker (1925).

Even if Greek philosophy is a philosophy of the Greek and for the Greek, yet the Greek was a man, and his city was a State; and the theory of the Greek and his polis is, in all its essentials, a theory of man and the State—a theory which is always true. The setting may be old-fashioned: the stone itself remains the same. We do not come to the study of the philosophy of the city-state as to a subject of historical interest: we come to the study of something, in which we still move and live. The city-state was different from the nation-state of today in the sense that it is a more vital and intense form of the same thing. In it the individual might realize himself more easily and clearly as a part of the State because its size permitted, and its system of primary government encouraged such realization. In studying it we are studying the ideal of our modern states: we are studying a thing, which is as much of today as of yesterday because it is, in its essentials, forever (p. 15).

Recalling ancient Greece we think of a landmass and offshore islands not much bigger than New England; its coastline, indented by numerous arms of the sea, is the home of Poseidon; the rugged mountains, the seat of gods; its chasms and caverns, the entrance to the underworld. By 700 B.C. the inhabitants of this area were the product of the amalgamation of the remnants of the Cretan, Mycenean, and Achean civilizations implementing the customs of the Dorian invaders and the Ionian settlers and incorporating some of the traditions of Egyptian and Phoenician societies. The ethnic diversity and the geographic topography of this area promoted the development of separate and distinctive city-states over a wide range of sociopolitical organizations, with Sparta and Athens at opposite ends. These city-states are taken as the initial model systems of sociopolitical complexity because they are relatively small units in which relationships between social classes are more transparent and the role of individuals in shaping society is more distinctive than in modern societies. The development of social organization preceded the philosophical thinking about it and is closer to human experience, and therefore suited for an initial examination of two levels of sociopolitical complexity and the meaning of conceptual continuity.

According to the classification given in the Introduction (table I.1), any society is a form of complex reality. Complexity refers here to the relationship of the individual and the social classes to one another and to the society and the state as a whole. The greater the specificity and diversity of the interactions and the less static the nature of these relationships at each level of organization, the more extensive the establishment of conceptual continuity. This is demonstrated by a comparison of the high-level and diminished complexities of Athenian and Spartan societies respectively.

The description of the specific forms of concrete sociopolitical complexity in the first part of this chapter is followed by a survey of some of the general, abstract simplifications of reality introduced by the philosophers of ancient Greece.

THE HIGH-LEVEL COMPLEXITY AND CONCEPTUAL
CONTINUITY OF ATHENS' DEMOCRATIC SOCIETY

Athens developed a form of governance in which at least a substantial minority of the total population were given freedom to pursue their lives as they saw fit and were actually encouraged to express their opinions on the affairs of the state. Beginning with the eighth century B.C., Athens became the capital of Attica's unified clans and townships. This unification, which was achieved more through persuasion and voluntary action than through force, was nearly complete by 700 B.C.

Athens' path toward the high-level complexity of governance started then with a conventional, authoritarian kingship. This was modified by the establishment of an archonship, similar to the office of a prime minister. Subsequently, the duties of a single archonship were distributed among nine archons, each of whom assumed responsibility for one of the administrative or legislative functions of the government, with the result that the power of the king was weakened. In that period the enfranchised citizenship included what was regarded as the aristocracy, its members belonging to long-established and well-known families of considerable wealth (consisting mainly of landholdings). The aristocrats lived in town in relative luxury; their slaves cultivated their surrounding lands. A second level of enfranchised citizens became a well-established middle class, including professionals, the businessmen who maintained the commerce of the city (this was possible in Athens but not in Sparta), craftsmen, laborers, and free peasants working their small holdings. A third social level, disenfranchised and numerically the majority, encompassed Athenian women and slaves.

Although during the initial phase of Athenian history some form of assembly existed that gave the middle class an opportunity to express their opinions for or against an item of legislation, the government was in effect controlled by the aristocracy, primarily through its economic power. Only during the later part of the seventh century B.C. did the middle class begin to accumulate sufficient wealth to exert some influence on the legislative process. Low-income citizens, in particular subsistence farmers, often owed money to rich citizens. The most pernicious aspect of this indebtedness was the rule that failure to pay back a loan would result in the enslavement of the debtor. Feuds arose between prominent families and clans vying for power. These were settled by the introduction of restrictive laws drawn up by Draco (seventh century B.C.). At the same time, the worsening economic condition of the poor class of free citizens increased tensions between the social strata of Athens. This polarization eventually became incompatible with the maintenance of the social fabric. An excerpt from Durant (1939) describes this predicament:

> As the seventh century drew to a close the bitterness of the helpless poor against the legally entrenched rich had brought Athens to the edge of revolution. Equality is unnatural; and where ability and subtlety are free, inequality must grow until it destroys itself in the indiscriminate power of social war; liberty and equality are not associates but enemies. The concentration of wealth begins by being inevitable and ends by being fatal. The disparity of fortune between the rich and the poor, says Plutarch, had reached its height, so that the city seemed to be in a truly dangerous condition, and no other means for freeing it from disturbances . . . seemed possible but a despotic power. The poor finding their situation worse with each year—the government and the army in the hands of their masters, and the corrupt courts deciding every issue against them—began to talk of a violent revolt, and a thorough redistribution of wealth. The rich, unable any longer to collect the debts legally due them, and angry at the challenge to their savings and their property, invoked ancient laws, and prepared to defend themselves by force against a mob that seemed to threaten not only property but all established order, all religion, and all civilization (p. 112).

In view of the dangerous situation in 594 B.C., the existing middle class commissioned Solon, one of the great political personalities, to mediate between the upper and the lower classes and to eliminate socioeconomic injustices as much as possible. The importance of the specific, nondeterministic accident in taking this first step toward democracy became apparent in Solon's appeal to virtue rather than force in achieving the decisive compromise.

As Sagan (1991) points out:

> In almost biblical language, he (Solon) gives us this paean to law and justice:
> "These are the lessons which my heart bids me to teach the Athenians, how that lawlessness brings innumerable ills to the state, but obedience to the law shows forth all things in order and harmony and at the same time sets shackles on the unjust. It smoothes what is rough, checks greed, dims arrogance, withers the opening blooms of ruinous folly, makes straight the crooked judgment, tames the deeds of insolence, puts a stop to the words of civil dissension, and ends the wrath of bitter strife. Under its rule all things among mankind are sane and wise."

Sagan comments further:

> In Rome no one with political power ever had this kind of vision of justice. In Israel, no one with this vision ever held political power. In Athens, miraculously, the vision and the power came together in one remarkable moment (pp. 55–56).

Solon relieved tensions by reducing debts and readjusting the currency value; he decreed the release of most political prisoners and eliminated the excesses of Draco's laws. He also introduced a form of progressive taxation and reorganized the government into a senate (a council representing the four original tribes), and an assembly open to all free citizens, who elected their representatives to the council and held them responsible for their actions under law. Even the members of the lowest income class could become jurors in the courts of the legal system. The citizens had access to a court of appeals where any unjust rulings of a magistrate could be rectified.

Under Solon's constitution, the archonships, the most prominently important offices, remained available only to the wealthy. After their one-year service as officials, the archons became members of an ancillary council, the Areopagus, that was established to oversee the Athenian system of laws (Ober 1989, p. 64). In contrast to the pre-Solon past, wealthy commoners, and not only aristocrats, could now aspire to these offices. In the assembly, free citizens of any economic or social rank were admitted and by majority vote could decide the current governmental agenda. The agenda, however, was prepared by a council of 400, which was probably still controlled by the wealthy.

Solon's reforms are emphasized here as an outstanding example of conceptual continuity in the sociopolitical sense. By identifying areas of common interest or by creating conditions that are of mutual advantage to two opposing social groups, a potentially dangerous increase in adversarial polarization is prevented and is replaced by the perception of a link between the two opposing social units. Such a link can be formalized by a common constitution, a treaty, or an agreement and is here designated as conceptual continuity.[1]

Solon's constitution provided sufficient continuity between the two classes of Athenian society, the elite and the commoners, to promote their coexistence without abandoning their identities. It created a society that existed not in a state of a static equilibrium but in a state of a dynamic disequilibrium and flexibility.[2]

Although the Athenian elite retained a great deal of power in managing the affairs of the state, Solon's constitution removed some of the blatant discrimination against the poor and prepared a path toward a greater cohesion of the Athenian citizenry and toward the recognition of specific areas of common interest in the goals of the separate classes with the concomitant establishment of sociopolitical conceptual continuity. Thus the constitution began to condition the Athenian citizen to accept the coexistence of the classes. It is the balancing of this coexistence throughout the remaining part of Athens' constitutional history that is its most pronounced complex characteristic. Solon's compromise in ancient Greece, one that adjusted the inequality of excessive wealth and poverty and that transformed an intolerable dichotomy into a tolerable duality, can be seen as a basic example for comprehensive democratic legislation. It has remained the main challenge to personalities with special appeal, talents, and skills to recognize specific conditions in establishing conceptual continuity in order to resolve crucial issues of democratic society.

After Solon left Athens for studies in Egypt (ca. 594–584 B.C.), shifts in the Athenian democratic constitution, the reversibility of the constitutional pendulum swinging from left to right and back again, reappeared. The conflict between aristocracy, middle class, and very poor had worsened once again, and it required dictatorial power, assumed by Pisistratus in 546 B.C., to reestablish order. Although of aristocratic origin, Pisistratus left Solon's constitution intact and actually reinforced the laws that had been decreed under it. He advanced the cultural life of Athens and elevated it to the first rank among the city-states. In view of these achievements, Pisistratus's rule can perhaps be regarded as one of the rare examples of a benevolent dictatorship (Durant 1939, p. 122; see also Sagan 1991).

After Pisistratus, Hippias, one of his sons, took the reins amidst intrigue and scandal. The disorders connected with his ascendancy compelled him to strengthen his position by reverting to the unsavory side of dictatorship: secret surveillance and terror. Invoking the help of Sparta against their own government for the first time, the citizens of Athens revolted, terminated Hippias's governance, and sent him into exile. Competition for the headship of Athens between Cleisthenes (leader of the revolt against Hippias) and Isagoras (favorite of the aristocracy) led again to an invasion by Sparta, this time on behalf of Isagoras; but the citizens of Athens successfully resisted the Spartan intervention, and Cleisthenes assumed power introducing at the same time considerable constitutional changes. He fully established the democratic form of government by enhancing the status of the assembly, giving it power to exile individuals who became a threat to the city-state's democratic institutions (ostracism). Cleisthenes deliberately fostered the civilian consciousness of the common man, thus replacing reverence for the elite by reverence for the state (Ober 1989, p. 67).

Cleisthenes further promoted the participation of the commoner in the decision-making process of the state. Citizens registered as voters in 139 centers *(demes),* comparable to our voter registration of today. The officials conducting the day-to-day business of the demes were elected irrespective of their social status; even the election of an aristocrat to a deme depended on the vote of commoners. Athenian council membership was increased to 500 and selection of council members was by plurality and removed from the influence of the elite.

After Pericles assumed the leadership of Athens (467–428 B.C.) even commoners gained full access to all governmental functions. Participation in the assemblies and expression of opinions in the assemblies, now held as many as forty times a year, were encouraged; property requirements for eligibility to higher offices were lowered; and most officeholders were entitled to some salary, provided by the state, making it financially possible for less affluent citizens to hold those positions.

With Pericles, leadership meant supreme skill as commander of the Athenian forces in the war and victory against Persia, brilliance as a public speaker, and a firm understanding of state finances. In his speeches he emphasized the achievement of the Athenian state and the desirability of imperial expansion. Moreover, he gave his rhetoric both concrete and symbolic meaning by the public constructions on the Acropolis. Under his building program, the addition of the Erechtheion to the existing temples there completed its full glory.

Athenian governance can be regarded as a model experiment.[3] For the first time in history, all freeborn native males were political equals under Solon's constitution and had equal rights and opportunities to debate and determine state policy. Debate was essential to recognize hidden factors that might affect the sociopolitical "steady state" order. This contrasts with the oligarchies that elsewhere remained the most common form of polis government (Ober 1989, p. 6). In Athens, any enfranchised citizen could speak and try to capture the attention of the audience during the decision-making meetings of the citizens' assemblies; these were attended by as many as 6,000 to 8,000 people (Ober 1989, p. 10). Because even Solon's marvel of democratic constitutional legislation did not outlast Solon's life, the democratic form of society had to be regenerated again and again, ever reasserting its vigor. Athenian society tolerated and even encouraged the coexistence of such sociopolitical alternatives as class stratification versus constitutional egalitarianism and freedom of speech versus the establishment of consensus between the elite and common citizen. This has been emphasized as an important trait of Athenian society (Ober 1989):

> While the evolution of the Athenian constitution may be seen as an attempt to employ political equality as a counterweight to social advantages, the Athenians

never achieved nor did they ever attempt a final constitutional resolution of the dissonance between the relative social and political standing of masses and elite. The Athenian masses were not willing to compromise the principle of political equality in ways that might satisfy the ambitions of the elite, nor were the members of the elite willing to part peacefully with the conditions of their superiority in order to alleviate the apprehensions of the masses (p. 295).

It seems that Athenians preferred to retain the flexibility together with the uncertainties of such complementary aspects of their sociopolitical organization rather than opt for a "fixed" solution by creating a one-sided dictatorship either of the masses or of the elites. In an artistic extension of this spirit, the great Athenian playwrights and their audiences indicated a deep awareness of the complexity of the human condition in the performances of the plays of Sophocles, Aeschylus, and Euripides, which explored the dramatic and tragic interactions of human beings with each other and with the gods.

With the increasing political influence and growing entrepreneurship of its middle class, Athens' economic role in the Mediterranean world acquired a prominence that began to threaten the vested interests of the Phoenician-Carthaginian Empire and that of Persia. Involvement in the rescue of its Ionic colonies in Asia Minor from Persian rule in 500–480 B.C. brought Athens into direct confrontation with that Near Eastern superpower. Through one of the rare cooperative efforts by several city-states, Athens emerged victorious, leaving for all time the battles of Marathon, Thermopylae, and Salamis as monuments to valor and heroism. Athens became a power with a dominant economic, cultural and military status in the world of the Mediterranean.

With all its promise, this Golden Age was only a short interval in the tumultuous history of the Hellenic states, for more often, the achievements of one of them became the envy of the others. Eventually the expansions of Athens into Italy and Asia Minor became a threat to Sparta that seemed to require a military solution. The resulting power struggle between Sparta and Athens weakened the resistance of the Hellenic edifice against the encroachment of the non-Hellenic superpowers—Persia and Phoenicia-Carthaginia—who were called in by one side or the other, sometimes with unexpected realignments. This intensified the internecine struggles among the Greeks and led to the use of the city-states in the political schemes of the superpowers. It is remarkable, however, that throughout this period the Athenian citizenry reasserted its democratic constitution even after the takeover of Greece by Phillip of Macedonia in 338 B.C., demonstrating the potential resilience of this complex form of society.

DIMINISHED COMPLEXITY AND CONCEPTUAL
DISCONTINUITY OF SPARTA'S TOTALITARIAN SOCIETY

Sparta was founded by Dorian invaders, who in about 1000 B.C. probably came in several waves from the north by land, over the Isthmus of Corinth. Other Dorian groups may have come by sea from other regions. Inveterate warriors equipped with superior iron weapons, the Dorians displaced or subjugated the indigenous Achaean population in the plains between the Parnon and Taigetos mountains in the Peloponnesus of southern Greece. The Dorians expanded their hold over a large part of the Peloponnesian peninsula during the first 200–300 years after their invasion, and by 630 B.C. they had vanquished the main opponents, the Messenians, after a long, drawn-out war. During this initial period Sparta was ruled by a dual kingship, assisted by a small number of overseers, the ephors, who might have acted somewhat as the members of a modern cabinet. It is not certain whether a separate legislative senate existed at this early time. The ephors may have represented the interests of families with a long line of well-established, landed predecessors who formed an aristocracy. Probably, the artisans, traders, and free peasants had some limited role in the decision-making process. A large class of serfs (five to six times the number of free citizens) were tilling the lands of the aristocracy.

The early period of Spartan military expansion coexisted with a flowering of culture: poetic creativity, advances in music and dance, and the development of the distinctive Dorian architectural style. Therefore it appears that military conquest and a good measure of cultural creativity were not mutually exclusive. This is vividly demonstrated again in the course of Athenian history. However, the freedom that gave Sparta its cultural self-expression existed only during a short initial period of Sparta's history and does not represent the Sparta we know from the conventional texts; we are more familiar with the later phase of Sparta's history. In the seventh century B.C. Lycurgus introduced a constitution according to which the state assumed complete control of the lives of its citizens, reducing individual diversity and, concomitantly, social complexity.[4]

The newborn male child was examined by a committee of experts who decided whether the baby was sufficiently vigorous to have promise for a military career. Disqualified children were thrown over a cliff or left to die from exposure in the Taigetos mountains (miraculous survivals were told as legends). The children who passed this brutal examination were reared by their parents until the age of seven. Thereafter they were enrolled in state institutions, mainly for vigorous physical training that would make them superior warriors. The intellectual part of this education was probably only minimal, for the curriculum emphasized subservience to the state rather than freedom of the individual as the supreme virtue, and it glorified victory or honorable death in battle as ultimate aims of existence. At twenty years of age the young men assumed obligatory military service while continuing their military training. They lived together in peer groups, sharing their meals in the barrack homes. Only at age thirty did the Spartan male who completed his training requirements acquire full citizenship. Up to that point in his life he was allowed to consummate any kind of homo- or heterosexual relationship, but marriage was expected only after reaching full citizenship. Failure to marry was regarded as a serious offense and procreation was a strict duty, to produce an adequate supply of future recruits. After age thirty and as an enfranchised, full citizen, the Spartan male joined an Assembly, with limited legislative significance. Membership in higher legislative bodies had an even higher age limit. Usually higher governmental duties would be assumed after retirement from military service at age sixty. This is the council of elders—the Gerousia. The Spartan male who did not participate in military training was barred from all social organizations and eked out a living as an artisan, trader, or free peasant. Entrepreneurship of any kind was restricted by law. Possession of gold or silver by anyone other than the state was prohibited and only iron could be used as currency.

In practice, the women of Sparta were the biological units in an eugenic production line that generated physically superior male warriors. Therefore, the Spartan women, too, were expected to achieve a correspondingly high level of physical fitness, exhibited in parades that included the Spartan womanhood in the nude. Their exalted biological role, however, gave the women in Sparta a higher social standing than those in other Greek city-states. But it is unlikely that Spartan women actually participated directly at any point in the political decision-making process.

The large majority of the Spartan population were slaves (helots) who tilled the lands of the aristocracy. Half of their produce was delivered to their masters. They could cultivate the land, marry and raise their families, in accordance with their own traditions and customs. Serfs were not allowed to leave their assigned plots of land and had to serve with their masters in wars and render any other needed services. The large population of helots was kept under control by strict supervision through an efficient system of secret police surveillance and insidious deceptions. According to Thucydides,

The helots were invited by a proclamation to pick out those of their number who claimed to have most distinguished themselves against the enemy in order that they might receive their freedom; the object being to test them, as it was thought that the first to claim their freedom would be the most high-spirited and the most apt to rebel. As many as two thousand were selected accordingly, who crowned themselves and went round the temples, rejoicing in their new freedom. The Spartans, however, soon afterwards did away with them and no one knew how each of them perished (Durant 1939, pp. 80–81).

It may be hard to understand why and how the citizens of Sparta submitted to the extreme totalitarianism of the Lycurgean constitution. Records of the proceedings of the Spartan ruling bodies are lacking, and no definite reason can be given for the imposition of a constitution with such monstrous impairments to individual freedom. With a citizenship of about 30,000 outnumbered by a helot population of more than 200,000, perhaps fear of a revolt by such an enormous numerical superiority may have justified the maintenance of a well-trained, permanently mobilized strike force. Any actual uprising of the helots was immediately and brutally crushed. As another possibility, the Spartan code may have been instituted in anticipation of a military showdown between the Dorian Spartans and the Ionian Athenians. However, Sparta subjugated the larger part of the Peloponnesus even before the imposition of its own constitution, without sacrificing the social diversity and cultural creativity of that period. After the introduction of its constitution, Sparta only completed its conquest of the Peloponnesus by subduing Tegea and Argos to become the leading military power in the Hellenic world. Quite likely this could have been accomplished under a less totalitarian constitution. Entering into a period of totalitarian rule in the sixth century, Sparta lost its internal cultural life and narrowly focused on its own interests. In its later history it actually sacrificed the Greek colonies in Asia Minor to obtain the support of the Persians in the war against Athens. In addition to the possible historical and sociopolitical roots of totalitarianism in ancient Greece, a paranoid condition has been suggested as a psychiatric interpretation. This paranoid condition is distinguished by fear and contempt for a real or fictitious adversary; in the case of Sparta, a fear of a possible revolt by the helots, of a failure of full control over the sources of this paranoid fear, and an aggressive stance against it, as for example the execution of those slaves with revolutionary potential (Sagan 1990, pp. 150–158).

The description of Sparta as a totalitarian society provides an opportunity to ask about the place of conceptual continuity in it. The existence of the individual is one-sidedly determined by general rules that are imposed by the state and restrict the expression of individual initiative toward social change, entrepreneurship, or artistic creativity. As far as any expressions of individuality exist, they are irrelevant for the maintenance of the state. If individual initiative in higher levels of organization is regarded as an important aspect of self-fulfillment and as the basis of sociopolitical conceptual continuity, it is practically nonexistent in the totalitarian state. The opposite alternative exists in the Athenian society.

The first century of post-Lycurgean Spartan history (600–544 B.C.) is discussed here because it is an example of a society of great stability. The sequence of ruling kings and their overseer ephors seems to be free of major interruptions, and if tensions existed between aristocracy, military, bureaucracy, and citizens, their resolutions must have been achieved without major destructive flare-ups; order apparently prevailed over centuries. Sparta's main aim was its superiority as a military power. Culminating with the triumph over Athens, a long series of victories punctuated by only a few defeats fulfilled this ambition.

In the context of this discussion, the organization of Spartan society is one of a low level of concrete, specific complexity. The term "simple" is avoided here because it is reserved for the definition of the abstract, ideal systems of physics, as suggested in the Introduction and in detail in Chapters 2 and 3. It was the prevalence of order in the Spartan state, however, that gave it a prominent place in subsequent discussions of abstract, simple reality particularly those in Plato's Academy, as considered later in this chapter. Although, when first approached, Sparta may give the appearance of a thoroughly virtuous society, with the commitment of its citizens to the highest integrity, loyalty to the state, and bravery in battle, it is actually a society of considerable ambiguity. Therefore, it is necessary to reappraise the views of those philosophers who glorified Sparta and to examine the demeaning price it chose to pay for its order. Durant says (1939, p. 87):

> Weary and fearful of the vulgarity and chaos of democracy, many Greek thinkers took refuge in an idolatry of Spartan order and law. They could afford to praise Sparta since they did not have to live in it. They did not feel at close range the selfishness, coldness, and cruelty of the Spartan character; they could not see from the select gentlemen whom they met, or the heroes whom they commemorated from afar, that the Spartan Code produced good soldiers and nothing more; that it made vigor of body a graceless brutality because it killed nearly all capacity for the things of the mind. With the triumph of the code the arts that had flourished before its establishment died a sudden death; we hear of no more poets, sculptors, or builders in Sparta

after 550 BC. Only choral dance and music remained, for there Spartan discipline could shine, and the individual could be lost in the mass. Excluded from commerce with the world, barred from travel, ignorant of science, the literature, and the philosophy of exuberantly growing Greece, the Spartans became a nation of excellent hoplites, with the mentality of a lifelong infantry man. Greek travelers marveled at a life so simple and unadorned, a franchise so jealously confined, a conservatism so tenacious of every custom and superstition, a courage so exalted and limited, so noble in character, so base in purpose, and so barren in result; while hardly a day's ride away, the Athenians were building, out of a thousand injustices and errors, a civilization broad in scope and yet intense in action, open to every new idea and eager for intercourse with the world, tolerant, varied, complex, luxurious, innovating, skeptical, imaginative, poetical, turbulent, free. It was a contrast that would color and almost delineate Greek history.

This reappraisal clearly juxtaposes the diminished complexity of Sparta and the high-level complexity of Athens. Eventually it was Athens' thrust toward the establishment of a Mediterranean empire that forced Sparta to transcend its virtual isolation and respond to the reality of a greater Hellenic world. In doing so, Sparta had to give up whatever comfort there was in maintaining its social fabric at a low level of complexity and had to assume the burden that came with involvement in the world beyond the boundaries of a single city-state.

This account of some of the elements of ancient Greek sociopolitical organization illustrates several aspects of complex reality with the range of diversity of governance and cultural achievements in two city-states. Within small geographic areas, we encounter the stability of the authoritarian rule of Sparta, in contrast to the fluidity of Athenian middle-class democracy alternating with authoritarian takeovers, quasi-feudal oligarchies, rebellions by the lower classes, and the simultaneous flowering of the arts as a unique contribution to posterity.

Spartan and Athenian societies give us just two levels of sociopolitical complexity. Spartan social organization, with its rigid conformity and persistence in its sociopolitical relationships, is seen here as representing a lower level of complexity. Athenian society stands for the higher level of complexity, with its less restrained citizenry, rapid changes in governmental structure, successive reassertions of quasi-democratic governance that involved readjustments between its social classes (an example of sociopolitical conceptual continuity), far-flung colonial commitments, and diversity of individual self-realization and artistic expression. The turbulent history of ancient Greece can be taken as a preliminary description of complex reality that must have been experienced by every Greek citizen, perhaps even by the slaves. In view of this, it appears remarkable, though possibly not unexpected, that in the midst of the complexity of almost constant political turmoil the first schemes for a simplified representation of nature were envisioned. Was it just this rapid flux of history that provoked some minds to substitute enduring and simple forms of reality for transience and complexity?

REALITY AS ABSTRACT SIMPLIFICATION: THE PRE-SOCRATIC PHILOSOPHERS

Ancient Greece did not accept the diversity of nature and the complexity of its city-states as unchallenged forms of reality. It was ancient Greece that originated the dichotomy of concrete, complex reality and its abstract, simplifying representation. The priests of Egypt and Babylonia accurately observed the courses of the planets and stars and predicted their locations and movements from season to season and year to year, but these observations did not lead to the construction of theories; complex and simple phenomena were not sharply differentiated, and any recognizable order was attributed to divine origin. It was also ancient Greece that, in the course of its complex sociopolitical development, introduced the abstract simplifications of reality as a characteristic of Western culture and as the origin of Western philosophical thought (Barnes 1965; Hussey 1971; Kirk and Raven 1984; Morris 1987).

Beginning in the sixth century B.C., the philosophers of the Greek cities in Asia Minor and Sicily sought answers to fundamental questions:

- Is the diversity of natural phenomena the result of the interaction of a limited set of a few basic elements?
- Is persistence or change the most general characteristic of reality?
- Is the structure of the cosmos, with its recurring changes, and the place of earth in it comprehensible in terms of some basic principles?
- Does the complexity of sensory perception give a misleading and confusing account of reality?
- Can the complexity of everyday life be replaced by a more general, simplifying form of reality?

The search for answers to these questions led to early attempts to achieve a simplification of complexity and to create a unifying and coherent representation of reality. These attempts may give the appearance of elementary constructs. They were, however, bold steps in initiating the development of three centuries of Greek thought that established the basis for the unfolding of the great philosophical systems during the subsequent phases of Western culture (Hussey 1971).

Thales of Miletus (Miletus was an Ionian city in Asia Minor) is usually regarded as the earliest (sixth century B.C.) Greek thinker. He was familiar with some of the basics of geometry, and his knowledge of astronomy is said by some to have led to the prediction of a solar eclipse in 585 B.C. In his view the universe was a hemisphere resting on water, with the earth floating as a flat disk inside at the base of the hemisphere. Thales postulated that the multiplicity of phenomena is but the manifestation of a single basic element (which had the consistency of a liquid), expanded throughout infinity. This elementary material was thought to be in rotary motion with vortices in it generating the cosmos, the planetary system, and the earth. Like water, this elementary material could turn into gaseous mist or into solid states like ice or the sediments of rivers. By assuming such different states, the basic element was transformed into the concrete phenomena of our experience, the perceptible objects. However, this element did not exhibit any one of the qualities of perceptible objects. It is an abstraction, a boundless entity, that is the source of ordered change in the cosmos and on earth, assuming the role of a divine force as a monotheistic replacement for the multiplicity of gods in Greek mythology (Hussey 1971, p. 16).

Following Thales, Anaximander and Anaximenes perpetuated the Miletean school of thought. Anaximander (ca. 610–546 B.C.) postulated that the cosmos is derived from a limitless, imperishable principle, the *apeiron,* that is omnipresent and gives rise to the basic elements of opposites, the hot and cold, the dry and wet. By sequential, but undefined, processes these basic elements produce fire, air, sea, and land, the heavenly bodies, the natural phenomena of the earth, and eventually life itself. In addition to these theoretical considerations Anaximander constructed a spherical model of the heavens with the earth placed at its center. The earth remained poised in its central location "because of its equal distance from all points of the celestial circumference" (Kahn 1972, 1:117), and the moon, sun, and stars were spaced in geometric proportions, refuting a mythological interpretation of the cosmos's origin.

Anaximenes (main activity ca. 545 B.C.) attributed to elemental air the role of the primeval and pervasive cosmic agent. Rarefaction of air generated fire; the condensation of air gave rise, in succession, to winds, clouds, water, and finally earth. The hot and the cold remained as additional attributes of the elements in association with the different physical states. According to Anaximenes the earth was thought to be a flat, broad, and shallow structure. He perceived the stars, sun, and moon as fiery bodies attached to a rotating hemispheric membrane. Extending throughout the universe, the primeval air element was exerting a constraint that maintained the regularity of the cosmic processes (Diamandopoulos 1972). With these ideas the Miletean school of thought introduced a simplifying system, consisting of a small number of elements, as the source of the diversity of natural phenomena. The change required in the successive transformations of the elements was merely implied in the absence of conceptual continuity.

In the following development of Greek thought, change and persistence became the two alternative basic characteristics of reality. A decisive step in this direction was taken by Heraclitus of Ephesus (main work ca. 500–460 B.C.; Stokes 1972, Hussey 1971). Heraclitus again postulated that pairs of opposite states, the dry and the moist, the hot and the cold, act in combination in generating four elements. The dry and the cold produce the element earth; moist and cold give water; moist and warm produce air; and the dry and the hot produce fire. Perceptible objects of everyday life were thought to be composed of differing proportions of the four elements. In Heraclitus's system, change is thought to be the result of a dynamic interaction between opposite states, designated as strife; positive interaction is required in generation; negative interaction results in disintegration. These interactions between opposites presuppose some limiting element. Metaphorically, Heraclitus refers to the strings of a lyre as connecting the opposites of the bow and the body of the lyre, suggesting an early form of conceptual continuity between opposites (bow/lyre).

In Heraclitus's view, change was the main characteristic of reality. Since fire was the element that combined matter and energy, it was the main driving element in maintaining continuous change. The dynamic nature of order and the role of fire in sustaining it is reflected in Heraclitus's famous words:

> The world order is eternal
> an everlasting fire
> kindled in measure
> and being extinguished in measure

Change occurs in a controlled fashion. Change in one direction is balanced by a proportional change in the opposite direction. The balance and order in this dynamic system is maintained by a universal principle, designated as Logos, that coordinates all specific events. As defined by Stokes (1972, p. 478), "The truth expressed in Logos is principally the unity of opposites. The opposites, however, are engaged in perpetual strife, and strife is not only right and normal, but is also, like the Logos, 'common' and everything happens accordingly." This may suggest an early hypothetical, general, and abstract form of conceptual continuity.

In Empedocles's (fifth century B.C.) work *On the Nature of Things,* written in metric form during his life in Sicily and other Greek colonies in southern Italy, earth, water, air, and fire become again the basic elements of nature. In the formation of the world the elements are first present in a homogeneous state of coalescence under the influence of attraction, which is followed by a differentiating separation of the elements under the influence of repulsion. Empedocles applied the principle of homogenization and differentiation, suggesting an early concept of evolution, to an interpretation of plants and animals. Attraction and repulsion became forms of interaction between the elements, emphasizing the distinctive dynamic quality of the four elements. For Empedocles, the physical categories of attraction and repulsion implied the anthropomorphic connotations of love and strife. In his *Purification,* Empedocles adopted these principles of love and strife as the basis for a religious system (Kahn 1972, 2:496).

The number of constituent elements was increased by Anaxagoras (ca. 500–428 B.C.). According to this philosopher, each piece of matter had an infinite number of potentialities for its actual manifestation, but the mechanism of the conversion of potentiality into reality remained obscure (Kerferd 1972, 1:115). As a final step in this development the ultimate constituents of matter were thought by Democritos (460–370 B.C.) to be indivisible, identical particles, the atoms. The properties of the atoms themselves did not change, but the assembly of the atoms in different proportions would generate a wide variety of observable properties.

Change, widely adopted by the pre-Socratic thinkers as the ultimate form of reality, did not remain unchallenged. Thus, Parmenides (born 515 B.C.) perceived change as "the way of seeming" and persistence as "the way of truth" (Furley 1972), anticipating Plato's theory of Forms.

As elementary as the pre-Socratic ideas may appear today, they are the first abstract representations of reality. These abstractions may be seen as the initial differentiation between the emerging Western modes of thought and the established ideas of Eastern philosophy (see Barnes 1965; Morris 1987, p. 211). Neither the mythologies of the East nor its empirical astronomy and mathematics involve rigorous, abstract, theory construction. It may be relevant that the absence of abstract thinking in China prevented for a long time the further development of its high level of empirical technology beyond the state it reached in the fourteenth century A.D. In contrast, the continuing expansion and refinement of abstraction in Western culture led to the unprecedented rise of theoretical and experimental science.

At this point, we can provisionally restate the meaning of complex and simple reality in early Greek philosophy. In one form of simplification, the diversities perceived in daily experience are modifications of a uniform, omnipresent material. This material has in itself the potentiality for the transformation of its parts into the different objects that surround us. A multitude of separate entities, the things observed, have been replaced here with a common material basis and a common transformation process. It is a conceptualization that is similar, in an elementary way, to Newtonian equations replacing the diversity of separate terrestrial and celestial forms of motion or to the substitution of quantum mechanical equations for the diversity of unrelated elements in the table of the periodic system. The decisive step common to all these schemes of thought is the one in which the initial complexity of sensory reality is replaced by the substitution of a form of imagined, simplified reality, independent of sensory perception.

The thoughts of these philosophers are an initial, daring attempt to avoid complexity by postulating that the diverse manifestations of nature are the products of the interactions of a small number of general elements. In an extreme view, these can be derived from a single, omnipresent element or agency. This foreshadows the relationships of the general properties of matter as conceived by the theories of modern physics. More specifically, one could see in these early ideas an analogy to the interconvertibility and conservation of energy of modern times. In this latter case the theoretical and mathematical formalization of these conversions, for example of work into heat, were complemented by the recognition of molecular motion as the link between these two entities, providing an independent interpretation of the work-heat relationship. As discussed in more detail in Chapter 3, the theories of modern physics gave rise, in this sense, to the establishment of conceptual continuity between the formally related entities. In contrast, the speculations of the early Greek thinkers did not yet lead to the identification of the links that established conceptual continuity between the related entities but led instead to various forms of hypothetical relationships.

During the following development the philosophical thought by Plato and Aristotle, the attempts at simplifying the perception of reality, were elaborated and expanded by aiming at an understanding of human existence.

THE SIMPLE AND THE COMPLEX IN THE APPROACHES
TO REALITY BY PLATO AND ARISTOTLE

The abstract simplifications of reality by the early Greek thinkers provided the basis for their subsequent extension by Plato and Aristotle that included considerations of ethical problems and proposals for the organization of ideal societies. Was it possible to reconstruct human existence from first principles or was it necessary to confront human existence as an irreducible form of complexity ? A native of Athens, Plato (427–347 B.C.) spent considerable time abroad, primarily studying mathematics. Eventually he settled in Athens, and in 387 B.C. he founded his Academy for the institutionalized development and propagation of his philosophy, with the help of Athenian citizens (Jowett 1920; Demos 1937; Ryle 1972). The Golden Age of Athens had passed by Plato's day, and the last of the Peloponnesian wars had ended Athens' leadership in the Hellenic world, heralding the decline of ancient Greece as political power. Although the disorder in Athens' democracy incurred Plato's wrath, it was that same disorder that permitted, within one city, the expression of opposing points of view, thus promoting Plato's Academy. Plato saw in Athens only the greed of its citizens, the excesses of liberty, and the instability of government. Yet, it was Athens where Plato found his haven. There was no place in Sparta for free thought, whereas it was tolerated, at least to a certain extent, in Athens.[5]

Teaching in Plato's Academy probably included mathematics, physics and metaphysics, history and literature, some form of music, and at the end of the curriculum moral and political philosophy. For Plato the complexity of sensory perception as a basis for the reconstruction of reality was suspect in all disciplines. The senses provided a confusing and ever-changing multiplicity of detail that seemed to defy systematic analysis—an analysis that could provide a coherent understanding of the universe. Only unchanging ideas, the products of thought alone, could lead to such an understanding. Accordingly, formalization of the universe in terms of algebra and geometry became the basis of Plato's approach, which led to the theory of Forms. Initially the abstract concepts in Plato's theory of Forms, as expounded in his *Phaedo* and in the utopian parts of the *Republic,* remained separate from sensory, empirical reality. Later, in the *Timaeus* and in the *Laws,* Plato demonstrated the relevance of his abstract ideas for the understanding of the concrete world.

In continuing the line of thought of some of his predecessors, Plato accepted the existence of the four elements, earth, water, air, and fire; but in Plato's scheme, the elements were subordinated to the primacy of geometric figures. In his *Timaeus,* Plato sets forth the argument that the elements are substantial, representing a certain level of solidity. Any form of solidity, in turn, is delimited by planes in three dimensional configurations. The cube is the geometric basis of the element earth, the icosahedron of water, the octahedron of air, and the pyramid of fire. The elements fire, air, and water can be transformed into each other. The states of the elements, such as the liquidity of water or its modification into wine, oil, honey, ice, hail, snow, or hoarfrost; or of fire as flames or amber; of air as mist or transparent ether are determined by the size of the elementary figures. All substances that melt at high temperatures, even gold, can be considered derivatives of water. There are also different kinds of earthenware, which are due to different mixtures of water and earth.

Plato's reduction of diversity to a limited number of geometric forms is possibly the most radical simplification of concrete reality. However, the link between the geometry of matter and its actual properties remained undefined. The geometric substitutes became isolated constructs without conceptual continuity to the other manifestations of existence. Plato tried to avoid this discontinuity by a further extension of his system of thought.

In accordance with his theory of Forms Plato makes a sharp distinction between "that which always is and is not becoming versus that which is always becoming and never is." That which is apprehended by our mind, intelligence, and reason is always in the same state; but that which is conceived by opinion with the help of sensation and without reason is always in a process of becoming and changing. To explain this, Plato introduces the idea of God, a supreme, single deity, as the origin of the eternal, unchanging ideas. The carrier of these eternal ideas, through which they become part of the human mind, is called the soul. The soul is created by God and is diffused through the material world and the entire universe. The physical elements—earth, water, air, fire—are initially present only in very small quantities and combine in adequate measure only after God has created the soul. Plato describes the relationship of the soul to the human body in some detail.

> The bones and flesh, and other similar parts of us, were made as follows. The first principle of all of them was the generation of the marrow. For the bonds of life that unite the soul with the body are made fast there, and they are the root and foundation of the human race. The marrow itself is created out of other materials: God took such of the primary triangles as were straight and smooth and were adapted by their perfection to produce fire and water, and air and earth—these, I say, he separated from their kinds, and mingling them in due proportions with one another, made the marrow out of them to be a universal seed of a whole race of mankind; and in this

seed he then planted and enclosed the souls, and in the original distribution gave to the marrow as many and various forms as the different kinds of soul were hereafter to receive. That which, like a field, was to receive the divine seed, he made round every way, and called that portion of the marrow, brain, intending that, when an animal was perfected, the vessel containing this substance should be the head; but that which was intended to contain the remaining and mortal part of the soul he distributed into figures at once round and elongated, and he called them all by the name 'marrow'; and to these, as to anchors, fastening the bonds of the whole soul, he proceeded to fashion around them the entire framework of our body, constructing for the marrow, first of all, a complete covering of bone (Jowett 1920, 2:51–52).

The idea of God was of great significance in Plato's thinking about human society, and disregard of God was the key to Plato's indictment of the moral decay of Athenian youth. Denial of God severs the self from the beyond-self and leads to a preoccupation with self-gratification: God created the individual for the sake of the whole, and not the whole for the sake of the individual (Jowett 1920, 2:645). For Plato, therefore, democracy was the least acceptable form of government. An insatiable desire for freedom leads to a disregard for law and order and opens society to being taken over by tyrants, motivated by their selfish drive for power and by their contempt for the rights of the citizens. In contrast, Plato's conception of reality, based on the primacy of abstract ideas, led him to a vision of an authoritarian utopia; a government that encompassed three social elements. The first element was a small ruling class of philosopher-statesmen with the philosopher-king having achieved the highest level of excellence among them. The members of this class have to submit to rigorous training in the basic disciplines, in two successive periods from ages twenty to thirty and thirty to thirty-five, reminding us of Sparta's disciplinarianism. As a rule, individuals who complete this curriculum are expected to join the ruling body after they reach the age of fifty (if their progress in the intervening years was satisfactory; only few would make it). The second elementary class of citizens was the military, which includes the police, who receive basic education from ten to twenty years of age, followed by specialized military training. Their basic education, comparable to, say, our secondary schools, begins with music for the mind and gymnastics for the body. In this context, music covers a broad discipline that includes some mathematics and an introduction to humanities, but very little instruction in what we would call natural science. All literature taught was to be censored to eliminate any works that might set a bad example. It was expected that this basic training would establish the proper balance of reason and passion. A third class consists of manual laborers with marginal education.

Under this form of government, diversity of conduct is rather narrowly limited and is rigidly controlled by the state. Sanctions are imposed directly by government officials of the philosopher-statesmen class in accordance with a concrete, detailed, and extensive list of laws (*Laws,* books V–X). Living and teaching in Athens and accepting its tolerance for diversity of thought, even criticism of democracy, Plato remained an advocate of the Spartan mode of government. It should be kept in mind that this is the first time that a reduction of reality to its abstract (geometric) elements goes hand in hand with a type of authoritarianism as a preferred form of government.

A second comprehensive system that reduced the complexity of reality was elaborated by Aristotle (384–322 B.C.) (McKeon 1941; Kerferd 1972; Guthrie 1981). Aristotle was a descendant of a long line of physicians and was exposed in his youth to a tradition of empirical observation of natural phenomena. When he came to Athens at age eighteen for further education, he acquired a foundation of philosophical thinking under Plato's influence. Although it is not certain whether he actually attended the Academy (Randall 1960), his early works bear the definite imprint of Platonic thought. After Plato's death in 347 B.C., Aristotle left Athens and spent several years traveling in Asia Minor. Much of his activity was devoted to the collection and examination of biological specimens, which became the basis for his great biological treatises. These were written after Aristotle's return to Athens, where he founded a school, the Lyceum, dedicated to philosophical contemplation.

Following his predecessors, Aristotle accounted for the different forms of matter by the admixture of the elements earth, water, fire, and air, in different proportions. (The composition of metals remained problematical, however.) The main direction of Aristotle's thought was to explain the relationship of matter, of any composition, to the generation and degradation of living and nonliving objects. For Aristotle, an inherent quality of the elements was their motion. Water and earth are endowed with a downward motion and air and fire with an upward motion. The existence of a separate force, or of interactions between bodies as sources of motion, was not recognized. The four elements and the two motions were the components of the terrestrial or sublunar sphere. The region beyond the sublunar sphere was the celestial sphere, which included the moon, sun, and the other planets and fixed stars. Their circular motions were attributed to a fifth element. A region of the prime mover, the source of all forms of motion, enclosed the celestial and the terrestrial spheres.

As the basis of existence, Aristotle's motion had a broad connotation; it was not just displacement in space but rather general change or process. In this sense, Aristotelian physics is an exploration of the functions of animated (living) and nonanimated (nonliving) natural bodies. Changes of the elements earth, water, air, fire, or of living matter, are the manifestation of an inherent quality of the changing natural bodies themselves. These changes are specific: earth going down, fire going up—it could not be the other way around—and all change has a purpose. Aristotle calls the power to change, or of acting in a certain way, the "nature" of an object. This is further clarified by the following quotations:

> Motion in space is not to be understood . . . as a succession of points occupied at successive instants of time. It is rather "the traversing of a distance." It is not a succession of determinations, but the determining of a succession, a continuous operation or process (Randall 1960, pp. 170–171).
>
> Nature is for Aristotle not an efficient cause: the "Nature" of a thing never itself does anything, any more than a physical "law" of nature ever does anything. The nature of a thing is its power of acting in a specific determinate way. . . . It implies . . . a structure to be investigated in the operating and functioning of that power (Randall 1960, p. 174).
>
> Motion (kinesis) is a process in which something which has the power to become a definite something else, becomes that something else. . . . It is thus the continuous actualization of what is potential, taken as being potential (Randall 1960, p. 178).

Aristotle developed a system of four causes as a generalized abstract explanation of the diversity of living and of nonliving matter. One favorite example for the identification of these causes is the elaboration of a candleholder from a piece of brass. The concept of the material from which an object is made, in this case the brass, is designated as the material cause. The effort for the transformation of the piece of metal into the candleholder, in terms of energy and processing required, is the efficient cause. In becoming a candleholder the piece of brass assumes a particular shape, which is determined by the formative cause. Lastly, the finished candleholder serves a definite purpose. It exists for the sake of holding a candle and giving light in a dark room as its final cause.

In accordance with this example, the definition of an object answers four questions: first, we ask what it is, for example, a flag. This answer is equivalent to the formal cause. We ask what it is made of. The flag is made of some coarse textile, the material cause. It was made by a specialized textile factory, the efficient cause; and it is used as a patriotic symbol, the final cause.

Aristotle's system is remarkable because his introduction of the four causes integrates the multitude of natural phenomena in the realm of planetary and terrestrial motion with the structure and function of living organisms. It was an early form of reduction that unified the complexity of perceived phenomena with a high degree of consistency. A clarification of the meaning of Aristotle's unification aids in understanding the relationship of unification and explanation in modern science, and Aristotle's interpretation of living organisms provides useful examples for examining his approach to unification.

It was, in particular, the manifoldness of living creatures as an important aspect of complex reality that Aristotle attempted to understand through the simplifying concepts of the four causes. In contrast to Plato, Aristotle did not start with preconceived ideas from which explanations of specific phenomena and objects could be deduced; he first carried out a systematic survey of living forms based on empirical observation *(Natural History of Animals)*. Only after completing the organization of this material did Aristotle proceed to a coherent explanatory account in terms of the four causes. Therefore, Aristotle's biology includes three separate approaches. One of these gives a systematic account of the main structural (anatomical) features of a great variety of animal species (curiously, plants are omitted from Aristotle's biological systematics). A second approach describes the generation of animals, comparable to modern embryology. The third approach, of primary importance here, gives Aristotle's account of the relationship of biological structure and function. The place of the four causes in this structure-function relationship is indicated in the following example, Aristotle's interpretation of kidney function (Ogle 1912, Section 671b).

> All animals whose lung contains blood are provided with kidneys. For nature uses these organs for two separate purposes, namely for the excretion of the residual fluid and to subserve the blood vessels, a channel leading to them from the great vessel. In the center of the kidney is a cavity of variable size. . . . The duct which runs to the kidney from the great vessel does not terminate in the central cavity, but is expended on the substance of the organ, so that there is no blood in the cavity, nor is any coagulum found there after death. A pair of stout ducts, void of blood, run, one from the cavity of each kidney, to the bladder; another duct, strong and continuous, leads into the kidneys from the aorta. The purpose of this arrangement is to allow the superfluous fluid to pass from the blood vessel into the kidney, and the resulting renal excretion to collect, by the percolation of the fluid through the solid substance of the organ, in its center, where as a general rule there is a cavity. . . . From the central cavity the fluid is discharged into the bladder by the ducts, that have been mentioned, having already assumed in great degree the character of excremental residue.

The bladder is as it were, moored to the kidney; for, as already has been stated it is attached to them by strong ducts. These then are the purposes for which the kidney exists, and such the functions of these organs.

This example shows that Aristotle is an accurate observer of anatomical detail. Moreover, he ascribes to each observed tissue component in the kidney a function by means of which it serves the organism as whole and which is its purpose, or its final cause. The material cause would be defined by the composition of the different components, the efficient cause by the process of percolation, and the formal cause by the shapes of the anatomical structures.

The four causes are concepts postulated by Aristotle to achieve consistency of thought about living organisms and about the diverse forms of matter in general. The introduction of the final cause as a link between biological structure and function remained unequaled as a unifying principle until the contemporary unification by conceptual continuity of structure and function at the molecular level.

In extending their lines of thought, both Plato and Aristotle devoted a large part of their work to an examination of society. As mentioned earlier, Plato begins the *Republic* by constructing an ideal form of society from first principles. It is a state in which the individual member of society is an entirely subordinated unit. In a subsequent treatise, the *Laws,* Plato deals with the concrete provisions that regulate the relationships of individuals and society. In contrast, Aristotle begins his *Nicomachean Ethics* with an empirical inquiry into the needs of individuals for an adequate life. Taking this as a starting point, he proceeds in his *Politics* with the development of comprehensive rules that define the place of the individual in society.[6]

Here Aristotle abandons much of the abstract conceptualization found in his other studies. An indication of a final cause remains in the opening section of the *Nicomachean Ethics,* when Aristotle asks about the nature of the highest good, the end toward which man is striving. That it is happiness seems at first an unexpectedly obvious answer. It becomes less commonplace when Aristotle inquires where happiness comes from and how it arises. Plato postulates an abstract, general goodness as the basis of all specific forms of happiness. In contrast, Aristotle recognizes multiple sources of happiness: the ordinary pleasures such as food or sex or, more refined, the beauty of nature or artistic creations. These are all experiences by and for the self. Pleasures that comes from political life, such as honors and civilian or military service, reward one for efforts made for the sake of the self as well as for the sake of others, or for society as a whole. The highest source of satisfaction is the life of contemplation, which ultimately is expected to contribute to the spiritual or material well-being of all of society (McKeon 1941, p. 1105).

For Aristotle, happiness is the result of what he calls "virtuous activities benefiting society and not just one's self." His account of virtue resembles an inventory of positive behavior with regard to others. This catalog of virtues includes temperance, virtues in dealing with money, honor, temper, social intercourse, and most importantly, justice. Happiness is activity in accordance with these virtues (McKeon 1941, p. 1102).

Simplifying generalization is problematical in Aristotle's interpretation of sociopolitical reality, and one can question whether a reduction of complexity is, or actually can be, achieved. Some difficulty is encountered in attempting to reconcile two definitions of the individual: on the one hand as a part of the whole of society and on the other hand as an entity unto itself. These two aspects are tentatively represented by the following quotations.

> Further, the state is by nature clearly prior to the family and to the individual, since the whole is of necessity prior to the part; for example, if the whole body be destroyed, there will be no foot or hand, except in an equivocal sense, as we might speak of a stone hand; for when destroyed the hand will be no better than that. But things are defined by their working and power; and we ought not to say that they are the same when they no longer have their proper quality, but only that they have the same name. The proof that the state is a creation of nature and prior to the individual is that the individual, when isolated, is not self-sufficing; and therefore he is like a part in relation to the whole. But he who is unable to live in society, or who has no need because he is sufficient for himself, must be either a beast or a god; he is not part of a state. A social instinct is implanted in all men by nature, and yet he who first founded the state was the greatest of benefactors. For man, when perfected is the best of animals but when separated from law and justice, he is the worst of all; since armed injustice is the more dangerous, and he is equipped at birth with arms, meant to be used by intelligence and virtue, which he may use for the worst ends. Wherefore, if he have not virtue, he is the most savage of animals, and the most full of lust and gluttony. But justice is the bond of men in states, for the administration of justice, which is the determination of what is just, is the principle of order in political society (McKeon 1941, pp. 1129–1130).

> Now the highest of all goods at which conduct can aim, all men agree, is the Good Life, or Acting Well, eu praxein: it is eudaimonia, that is "well being" or "welfare." The end of conduct is human welfare. . . . It is the one principle in terms of which all conduct is to be understood and judged. Now each kind of living thing has its own good, relative to its own specific nature. There is thus one good for fish, and another

good for birds, and still another for men. And the specific good for men, say, is likewise always "relative" to the particular man facing his particular problems, choices, and situations. There is no good common to all living beings; the good for fish is not the good for birds, and it neither is or could be the good of men. Likewise, there is no human good that is or could be common to all men on all occasions, save that of always acting intelligently. What is good is always something plural, specific, and relative to a particular situation or context. . . . But that does not mean for Aristotle that the good of any particular situation is "subjective" or "personal" or dependent on the mere "feelings" of the participants in the situation. It does not at all mean that statements about the several goods of different situations are not objectively determinable and discoverable, that such statements are not "scientific," but are merely "emotive," expressing the subjective "feelings" of men, or expressing the "existential commitments." Each situation has a good which intelligent inquiry can hope to discover. Aristotle is in ethics a complete and thoroughgoing relativist—an objective relativist, in our present-day classifications (Randall 1960, pp. 251–252).

From the first quotation it might be inferred that in Aristotle's scheme the subordination of the individual is as authoritarian as in Sparta's constitution and in Plato's ideal society. The second quotation, however, suggests a definition of the individual without direct reference to a relationship with the superimposed state. In resolving this dichotomy, Aristotle suggests that the larger social unit is essential for the maintenance of the subordinate smaller social units. In this sense the state, as the most inclusive social unit, is essential for the maintenance of society. The state's organization is required to provide the necessities for the largest number of individuals. But what form of government seems most adequate to allow for diversity of its components and still maintain the fabric of the state? Aristotle rejects the uniformity of the Platonic society with its communal state of women, children, and property. He also rejects the extremes of democracies in which only the poor are the ruling body. Surveying the constitutions of Sparta, Crete, and Carthage, he regards among those contemporaneous constitutions the conservative democracy of Solon as the best. Aristotle preferred a state based on a well-developed middle class and gave the following argument in support:

Now in all states there are three elements: one class is very rich, another very poor and a third is a mean. It is admitted that moderation and the mean are best and therefore it will clearly be best to possess the gifts of fortune in moderation; for in that condition of life men are most ready to follow rational principle. But he who greatly excels in beauty, strength, birth, or wealth, or on the other hand who is very poor, or very weak, or very much disgraced, finds it difficult to follow rational principles. Of these two the one sort grow into violent and great criminals, the others into rogues and petty rascals. . . . Again the middle class is least likely to shrink from rule, or to be overambitious for it; both of which are injurious to the state. . . . Wherefore the city which is composed of middle class citizens is necessarily best constituted in respect to the elements of which we say the fabric of the state naturally consists. And this is the class of citizens which is most secure in a state, for they do not, like the poor, covet their neighbors' goods; nor do others covet theirs, as the poor covet the goods of the rich; and as they neither plot against others, nor are themselves plotted against, they pass through life safely. . . . And democracies are safer and more permanent than oligarchies, because they have a middle class which is more numerous and has a greater share in the government (McKeon 1941, pp. 1220–1221).

Aristotle does not suggest absolute norms that restrict the individual to a minimal range for self-fulfillment. To him the existence of societal norms rather indicates the consensus that most efficiently provides the means for the largest possible number of citizens to attain the highest degree of self-fulfillment. The gain of belonging to a society consists in recognizing a double life experience in a single lifetime; of being an entity who is individual in one's self while being part of the greater whole of society. To become aware of this duality, to enjoy the burdens and pleasures of this dual existence in this higher level of complexity is probably one of the main purposes of education, and it may be regarded as the beginning of conceptual continuity in sociopolitics, which will be reconsidered in Chapters 2 and 5.

The survey of ancient Greece is here included because it juxtaposes two perceptions of reality that are examined throughout this book: the concrete and specific high-level complexities of the Athenian and Spartan city-states, and the abstract generalizations of reality set forth by the Greek thinkers. The high-level complexity of Athenian society is indicated, during its periods of democratic governance, by the freedom of expression of its citizens and the coexistence of its social classes. It was Solon who introduced a constitution that promoted the creation of areas of common interest between contending classes and established in this sense conceptual continuity as a basis for political action and understanding under conditions of sociopolitical diversity. In spite of rapid shifts in its form of governance, all the way from democracy to benevolent dictatorship to outright tyranny, the Athenian citizenry reasserted again and again its democratic rights.

The spirit of Athenian complexity found unparalleled expression in its contributions to Western culture, with its temples and amphitheaters, in which dramas and tragedies bared the last recesses of the human soul, and with its sculptures that portrayed the beauty of the human body and the dignity or the wrath of the gods. Even though Athenian democracy excluded women and slaves from suffrage and was lacking in the stabilizing division of power and checks and balances, the dynamic state of Athenian democracy marks it as a sociopolitical prototype of high-level complexity.

Sparta's authoritarian form of government eliminated diversity and flexibility and maintained a uniform social structure of apparently diminished complexity over centuries. In the absence of diversity there was no need for a search to identify areas of common interest between contending classes; the only area of interest was the state as a whole, without an opportunity for the establishment of conceptual continuity between diverse sociopolitical elements. Athenian complexity engendered individual freedom and diversity with all its dangers and its burdens of responsibility. Spartan diminished complexity removed these burdens but quenched individual freedom at the same time. It foreshadowed the curtailment of complexity and of conceptual continuity by authoritarian and totalitarian forms of nationalism and fundamentalism.

In attempting to reduce the pervasive experience of complexity, the early Greek thinkers constructed a simplified form of reality as the result of the interactions of a small number of common elements. Although the nature of these interactions was purely speculative and conceptual continuity cannot be imputed to this early form of simplification, it was a decisive step toward the unification of physical reality traced in the following chapters.

During a later phase of Greek philosophy, Plato and Aristotle, starting from first principles, reconstructed ideal simplifications of human society in the *Republic* and *Nicomachean Ethics,* respectively, suggesting in the corresponding *Laws* and *Politics* the concrete steps for the governance of the complex diversity of social organization. This raises the question of the relevance of general principles and theories for the understanding of the highly specific interactions of the components of complex inanimate, living, or social systems. The search for an answer to this question is one of the main themes pursued in this book.

Chapter 2 The Reassertion of the Simple-Complex Dichotomy in the Modern Era

Following the last phase of Hellenic dominance of Western culture (323–30 B.C.) a high-level complexity in everyday reality must have persisted. The next several centuries saw the ascendance and collapse of the Roman Empire, the invasions of the Mediterranean world by northern tribes, and struggles between the Christian and Islamic religions. Whether the extension of the early Greek philosophies in the Epicurean and Stoic systems of thought formed tranquil islands of simplicity is left unanswered here. Only after this historical tumult did individuals in Western society perhaps find a refuge from the complexity and trials of the daily routine in the dominance of the Christian faith (600–1200 A.D.), with its divine order of the universe and the certainty of a celestial hereafter as reward for a virtuous life on earth. This order was manifest on earth in absolute monarchies, with their hereditary continuity, the emperor being crowned by the pope to represent the secular arm of an ordained justice. A social order firmly grounded in aristocracy and feudalism did not encourage the stress and strain of social mobility and instead gave the comfort of static permanence between the authoritarianisms of secular, absolutist tradition and ecclesiastical doctrine. There was only a minimal leeway for abstract sociopolitical schemes and expressions of individuality. Perhaps during these times restriction of individual self-expression was a worthwhile price to pay for peace of soul and mind.

The changes in the state of human existence following 1200 A.D. have fascinated many scholars interested in the history of science and of culture in general (Butterfield 1950; Barnes 1941, 1965). A multiplicity of factors promoted a basic reorientation of society. The rise of a middle class of merchants and entrepreneurs (guilds of craftsmen) coincided with the far-flung expeditions that brought the rest of the world to European societies. The flow of gold and silver from newly conquered colonies and the arrival of new commodities, spices, textiles, and other products of foreign arts and crafts now affected the everyday life of a small elite. The broader availability of the treatises of the ancient Athenian thinkers contributed to the reassessment of social values. The precepts of a rising individualistic humanism once again engendered the perception of a complex reality. The new trends toward religious reform that led to prolonged religious wars diminished the unquestioned, monolithic security of the church. This fundamental change in attitude becomes strikingly manifest on comparing the representation of reality given by St. Thomas Aquinas (1224–1274) with that given by Hume (1711–1776). According to Aquinas, all forms of existence express diminutive, but distinctive, aspects of eternal law; expressed through human reason, eternal laws constitute natural law.[1] The highest forms of human thought were but remote reflections of God's mind, however, and governance was essentially a matter of divine providence.

In contrast, Hume, metaphorically summarizing the view of the last phase of the Enlightenment, presents the world, reality as a whole, as an assembly of machines. The mind of God must be mechanistic, although on a vastly grander scale than the mind of man.[2]

In antiquity the simplifying representations of reality were part of speculative philosophical schemes. Now, with the onset of the modern era, the developing disciplines of science, in particular physics, took the place of pure speculation. A part of the scientific endeavor aimed at the practical improvement of human existence by describing, collecting, and cataloging animals and plants, and by aiding agriculture and manufacturing (Bacon, for example, 1561–1626; see Anderson 1948). These studies increased the awareness of the diversity and complexity of nature.

At the same time the quest for a secular simplification of reality became increasingly prominent. The initial phase of this development, the Copernican revolution, was followed by the emergence of Galilean-Newtonian mechanics. During this time man's vision and reasoning became adjusted to the reality of a heliocentric planetary system, and the postulated circular motion of the planets was replaced by a mathematical representation of their elliptic paths derived from actual observation. The idea of "force" became the center of mechanistic thinking and a step toward conceptual continuity. Force as interaction between bodies over a distance became the capstone of the new construct of the planetary system.

Acknowledging the great influence of these developments on the thinking of the Enlightenment, a brief account is included here of the main steps leading to and through the two hundred years from Copernicus (1473–1543) to Newton (1642–1727) and of the mechanistic system of thought. Although in general well known, this transition is briefly reviewed as an essential prerequisite for the interpretations of some of the Enlightenment's main lines of thought in the subsequent sections of this chapter (Figure 2.1).

THE EMERGENCE OF THE IDEAL SYSTEMS OF PHYSICS:
THE GALILEAN-NEWTONIAN MECHANICS

Prior to the studies of Copernicus, the proposed schemes for the motion of the planets and stars were based on Plato's idea of the geometric perfection of the universe as elaborated by his pupil Eudoxus (360 B.C.): transparent crystalline spheres were the carriers of the celestial bodies in their circular motions about the earth. Actually, Pythagoras (582–500 B.C.) had postulated earlier that the ratios of the planetary distances from the earth should correspond to the ratios of the harmonies produced on a stringed instrument. As observations on the motions of the planets accumulated, an increasing number of modifications had to be added to the original single, circular orbits to account for the actually observed planetary paths. Eventually, the most elaborate scheme for the movement of celestial bodies was developed by Ptolemy (ca. 130 A.D.). In his system the planets follow a doubly circular path, a large-diameter circle around the earth with small-diameter circular progressions (epicycles) spiraling around the large circle. Ptolemy summarized his interpretation of the celestial movements in a treatise that became famous under the title of its Arabic translation, *Almagest* (The Greatest Work), the form in which it was preserved (as were many other manuscripts of antiquity). The data in this treatise were of an accuracy that made them useful guides for many centuries, but the complexity of this geocentric system seemed incompatible with the harmonious simplicity

Figure 2.1 Temporal Relationships Among the Great Thinkers of the Enlightenment.

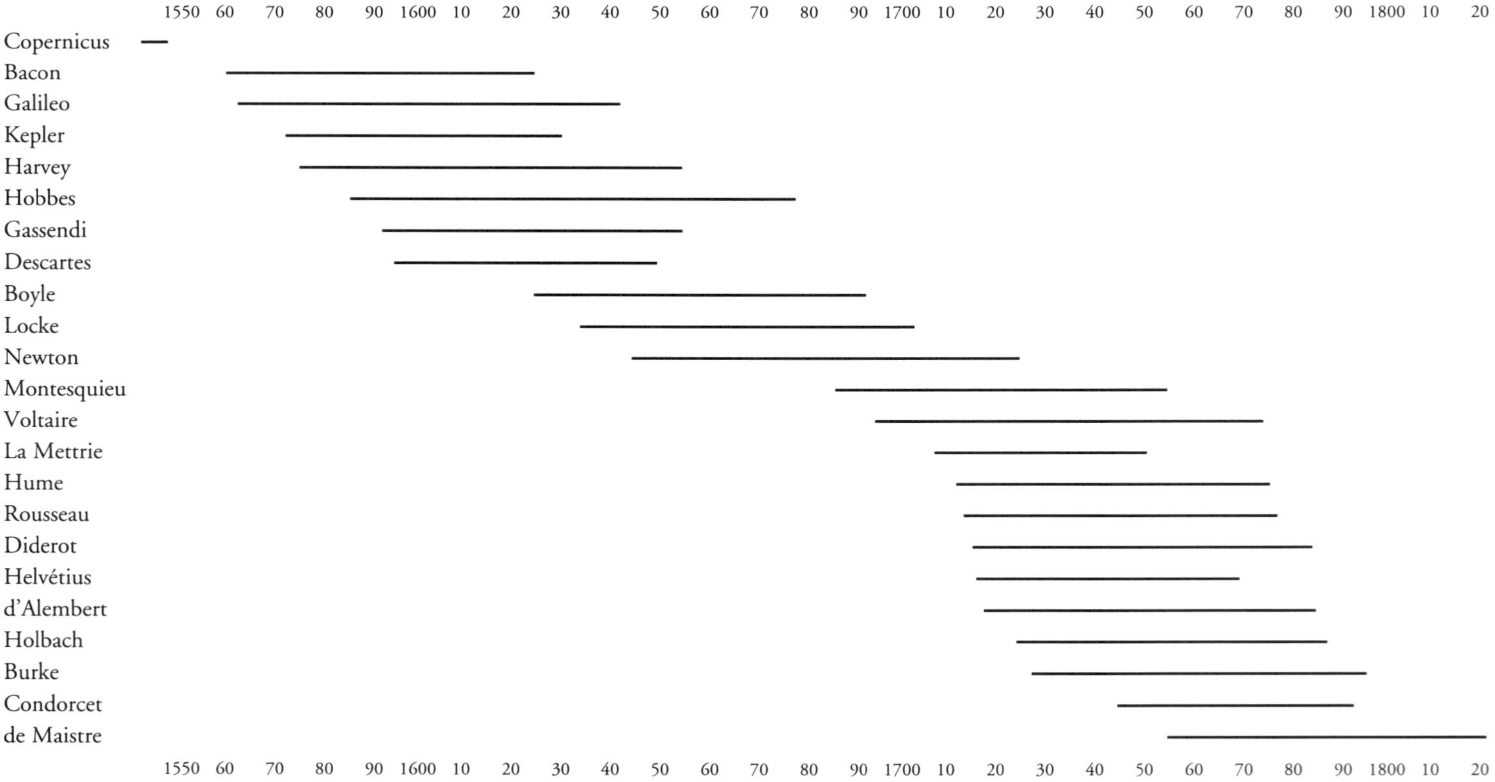

expected as one of the characteristics of a universe designed according to the prevalent religious ideas. Attempts to find simpler solutions, including heliocentric schemes, were made sporadically during the centuries following the original appearance of Ptolemy's work. It was Copernicus, however, who carried through the systematic effort of matching the available observations to the formal requirements of a heliocentric system. His heliocentric scheme led to a much simpler representation of the planetary motions; in particular, it accounted for planetary retrogressions as a function of the earth's movement. Yet even Copernicus retained some of the complicating epicycles, the circular courses of the planetary paths, and the organization of the universe into crystalline spheres as underlying vestiges of the Ptolemaic scheme. The religious and intellectual tolerance of Copernicus's time and place permitted the acceptance of a heliocentric scheme, but only a century later an intolerant church might have forced Copernicus to either recant or face execution (Ravetz 1990).

With the sun now placed center stage in Copernicus's scheme of the planetary system, the earth became merely another one of the seven known planets, changing the vantage point for the reappraisal of humankind's place in the universe. From a humanist view, Copernicus's achievement was indeed of revolutionary significance (Kuhn 1957; Cohen 1976). But since it relied on ancient concepts it contributed nothing to the question of the mechanism that keeps the celestial bodies moving in their unchanging paths. From a purely scientific point of view Copernicus's reconstruction of the planetary system was only a first step (Phase I, see legend to Table 2.1) toward a more satisfactory

---

Figure 2.1 is a survey of the temporal relationships among the great personalities who contributed seminal ideas to the development of thought from the end of the sixteenth through the seventeenth and eighteenth centuries. The time span that began with the achievements of Copernicus, Galileo, and Kepler and ended with the publication of Newton's *Principia* in 1687 preceded or coincided with the creative periods in the lives of Descartes, Hobbes, and Locke. The French Philosophes, Montesquieu, Rousseau, Burke, and de Maistre did their main work after Newton's crucial contribution to the full establishment of mechanics.

Descartes, Hobbes, and Locke emancipated human thought from the thrall of extra-natural doctrines and certain secular traditions, and instead adopted mechanistic concepts for the understanding of human reality. The French Philosophes, in particular Holbach and La Mettrie, developed the mechanistic interpretation to its extreme. In contrast, Montesquieu, Rousseau, Burke, and de Maistre rejected the mechanistic simplifying interpretation of reality.

Table 2.1 Phases of Scientific Analysis Leading to Conceptual Continuity
in the Ideal System of Physics

	Proposed *identification*	Conventional *designation*
Phase I	Recognition and characterization of specific phenomena	Descriptions
Phase II	Establishment of quantitative correlations between phenomena	Laws
Phase III	Introduction of conceptual relationships	Theories
Phase IV	Establishment of conceptual continuity	Explanation

The main steps in the analysis of physical phenomena are summarized here as an introduction to the scientific basis of the Enlightenment. One can suggest four phases in this process. It begins (Phase I) with direct observations and qualitative descriptions of previously unrecorded phenomena (e.g., electric repulsion and attraction) or with a qualitatively different interpretation of existing information (e.g., heliocentricity of the planetary system). In Phase II, the qualitative observations are restated in a quantitative form as mathematical correlations, such as Kepler's laws. The magnitudes in these quantitative correlations remain conceptually unrelated at first (e.g., mass and motion, mass and electric attraction and repulsion). Phase III is a further development in which concepts are introduced (e.g., force, field, wave propagation) to establish formal relationships (theories) between the initially unrelated magnitudes. However, a conceptual gap can remain between the related entities. For example, a relationship of matter, force, and motion in Newton's equations is taken for granted. Even today, however, the actual nature of the extremely weak gravitational force and its interaction with matter is still based on conjecture. It should be reemphasized that in the understanding of physical reality Phase III was of immense practical importance in the industrial development of the West since energy production and use and the mechanics of engine operation are all based on the theoretical advances made in this phase (nineteenth-century physics). Phase IV in the progress of scientific analysis consists of the establishment of conceptual continuity between the formally related variables. This step is a form of scientific explanation and is discussed in Chapters 3 and 4. A survey of the established theories of scientific explanation is provided by Pitt (1988). One of the included interpretations of explanation (Kitcher 1988) suggests some relationship to conceptual continuity.

understanding of planetary motion, but it was invaluable in preparing the way toward that goal.

Not until after the death of Copernicus in 1543 did the extensive and painstaking measurements by Tycho Brahe (1546–1601) provide an adequate basis for a quantitative and accurate definition of planetary motions (Phase II). Using Brahe's data Kepler (1571–1630) first tried to coordinate the allegedly circular planetary paths to the five geometrically perfect solids held over from the Platonic utopia of the universe. Attempts to calculate the path of Mars as a circular orbit, in particular, did not agree with Tycho Brahe's very precise measurements. Having found a discrepancy of only 8 minutes of arc—an almost negligible discrepancy for those days—Kepler insisted on the significance of this relatively small deviation that forced him to relinquish the firmly entrenched idea of the perfect circular motions of the planets. Instead, Kepler demonstrated that elliptical paths are in complete agreement with Brahe's data and that, on this basis, the motions of the planets can be fully defined by three laws (Blanpied 1969, p. 73):

Law 1. The orbit of a planet is an ellipse with the sun at one focus.

Law 2. The radius vector of a planet (the line connecting the sun and the planet) sweeps out equal areas in equal times.

Law 3. The square of the period of revolution of a planet about the sun is directly proportional to the cube of the radius vector of its semimajor axis. (The period of a planet is its "year"—the length of time it requires to complete one revolution about the sun. Its semimajor axis is one half of the greater dimension of its elliptical orbit.)

Compared to preceding treatments of planetary motion, the mathematical sophistication of these laws is striking, considering that the elliptical orbit deviates from the previously accepted circular orbit by only 0.5 percent (Blanpied 1969, p. 75). Planetary motion and planetary matter still remained conceptually unrelated, however; there is no indication of a link between the two concepts, that is, no mechanism suggesting the underlying reasons why the planets were being held in their orbits and hence no real conceptual continuity between matter and motion. In this sense we are dealing with precise, quantitative, but merely descriptive correlations. In his own attempt to account for the planetary orbits, Kepler suggests that spokes extending from the sun propel the planets into their elliptic courses, an entirely hypothetical connection, incompatible with Kepler's own laws.

Just as Copernicus did before him, Kepler's demonstration of the ellipticity of planetary motion abolished another component of the ancient astronomical ideology, circularity, as evidence for the perfection of nature. Although Kepler himself did not arrive at a valid mechanism that explained planetary motion itself, his laws provided the sort of information that invited the addition of the missing steps, the explanatory stones necessary to complete the observational and descriptive edifice. The first of those steps was the great contribution of Newton. That mathematical formalization by itself does not necessarily explain the mechanism of a process such as planetary motion is corroborated by the following evaluation of Kepler's laws (Blanpied 1969):

> Although the three laws of Kepler completely replace the complex tables of deferents and epicycles required in the Ptolemaic and Copernican descriptions, revealing a simplicity and harmony that the earlier systems had succeeded in almost completely concealing, the laws are still purely kinematical (descriptive). They go no further in themselves toward suggesting *why* the planets move as they do and in that sense are still bound to the old spirit of astronomy as a purely mathematical exercise. But Kepler was not content with pure kinematics; all of his career had been motivated by a desire to understand the reasons for planetary motion. The simplicity manifested in his mathematical results strongly suggests the possibility of finding a basic physical mechanism underlying the whole system, thus providing a connection between physics and astronomy (p. 83).[3]

Kepler's laws dispose of the Ptolemaic complexities of planetary motion; from now on, epicycles and hypocycles are banished. More significantly, the simplicity of the relationships Kepler formalized suggests the existence of some even more basic, explanatory principles which are not directly evident in the laws. It is not just the harmony of mathematical expressions but their role as guides to the why of physical processes that gives them such importance. We will return to this caveat about the relevance of mathematical representation in the discussions of the mathematical treatment of complex, for example, biological systems.

THE FORCE CONCEPT AND THE DEFINITION OF
THE IDEAL SYSTEMS OF PHYSICS

One of the decisive steps beyond Kepler's descriptive mathematical expression of planetary motion was the experimental approach by Galileo Galilei (1564–1642), which led to the concept of force as defined later by Newton. As opposed to Plato's belief, Galileo asserted that the most general properties of material objects could neither be established purely by speculation nor deductively only by reason. Instead, most general properties had to be established experimentally by comparing a large number of different systems under different conditions. In these experimental models, the dissimilarities among diverse systems are eliminated, and only their similarities are recognized as the most important representations of the reality of those systems. This point of view differs sharply not only from Plato's purely conceptual reconstruction of reality, but it also differs from Aristotle's approach, in which the negation of any particular attribute of a system may affect the state of the entire system (see Blanpied 1969, p. 4).

In following his experimental approach to gravitation, Galileo measured the time required for spherical bodies to roll down surfaces tilted at different angles. He observed that the rate of acceleration of the bodies was the same in each case, and by extrapolating this information to the vertical he could approximate the acceleration for free falling bodies. The conception of a single, general aspect of reality, the constancy of acceleration, of all bodies was substituted for the diversity of objects used for the tests. Galileo established one of the most general characteristics of any form of motion and gave it a mathematical formalization. He did not take, however, the next step, the establishment of the concept of force as the basis of acceleration. Thus, the relationship of Galileo's acceleration of motion to Kepler's elliptical orbits of the planets remained obscure. Newton's pivotal accomplishment was the recognition of the same relationship between mass, motion, and force for terrestrial and celestial bodies as one of the fundamental simplifications of reality.

Two different phenomena are frequently combined in discussing the concept of force. The first of these is force as observed in direct-contact interactions between solid bodies, the well-known example of billiard balls. The second phenomenon is interaction between masses at a distance, gravitational force, for example, which is inherent in all forms of matter.

Newton defined force in the form of three laws. The first law states:

(1) *A body continues in a state of rest or in a state of uniform, linear motion until acted upon by an external force.* Accordingly, linear velocity, that is, motion without change of speed or direction, is the same as vector velocity in a later form of the first law. The velocity vector of a body remains constant in a force-free state (Blanpied 1969, p. 84). In this form the first law can be said to assert that the action of force is defined by an increase in the velocity or a deviation from linearity of motion. Continuing action by a force produces an increasing acceleration of linear motion or a persistent deviation from linearity. The elliptical paths of the planets or acceleration during free fall are manifestations of the pervading gravitational force of the sun or the earth respectively. This leads to Newton's second law for accelerated motion.

(2) *The total instantaneous force (F) applied to a body is equal to the mass of that body (M) times its acceleration (a) (F = Ma).* Sufficiently sensitive apparatus can demonstrate that the force of gravitational attraction between two

bodies is directly proportional to the product of their masses and inversely proportional to the square of their distance from each other. The mutual interaction between two or more bodies is expressed as Newton's third law.

(3) *If a body A exerts a force FA upon a body B, then the body B exerts an equal and opposite force FB on a body A, such that FA − FB = o (Blanpied 1969, p. 106).* Newton introduced differential calculus as a mathematical formalization of the relationship of mass, force, and motion, as a conceptually unifying interpretation of celestial and terrestrial forms of motion, with gravitational force as a basic, unifying form of mechanical interaction over distances in space.

In the Newtonian systems, gravitational interactions, from the legendary falling apple to the planets in their elliptical paths around the sun, are defined by the same magnitudes: mass, force, and what are called the state variables, such as locations in space at successive times to represent motion. The precise mathematical definition and experimental confirmation of gravitational interactions gave the Newtonian concept of force an objective reality missing from Aristotle's concepts of the unmoved mover and final cause and from the solar spokes Kepler used as force-analogues in the interaction between bodies in general and between planets in particular. Newtonian mechanics proved adequate for predicting the existence of then-undiscovered planets and even in charting today's space travel. The planetary system was thought to be fully defined by these magnitudes and their mathematical relationships. No other factors besides force and mass are considered in defining the state of the system or in predicting its changes. Thus, in calculating the acceleration of the falling apple friction with air is ignored. The minimal gravitational effects of asteroids cruising through the planetary system on the daily or annual revolution of the earth can be disregarded. By choosing gravitational attraction as the only general characteristic of interacting bodies, any of the specific properties of the interacting bodies became irrelevant. We now have a conceptual construct designated as an ideal system (see Table I.1). Our usual idea of physics represents reality in the form of such ideal systems, arrived at by extensive abstraction. Defining such systems only by their most general characteristics simplifies reality. Change from one state to another in such systems must be predictable with certainty, with a small number of known variables determining the state of such systems. In this sense ideal systems can be termed fully determined or fully defined.

The Newtonian representation of reality is of central importance in this chapter because it gives meaning to the terms unification, simplification, and ideal systems. By relating the most general properties, mass and motion, of all concrete objects, through the concept of force, diversity is replaced by uniformity, and complexity by simplicity. Also, by introducing force as a conceptual link between mass and motion the Newtonian system is fully defined. It should be recognized, however, that the introduction of force did not provide real conceptual continuity for all systems. Such continuity exists only in systems with a direct energy transfer from one moving body to another, as in the familiar billiard-ball system. Conceptual continuity was missing in gravitational interactions at a distance. This became a basic concern for Newton himself, since he could not conceive of forces interacting between masses at a distance without some material transmitter. Newton felt compelled to attribute the role of such a transmitter to an elusive form of matter, a conceptual crutch since antiquity, called "ether," thought to pervade the universe. Although the existence of such an ether was later refuted, its initial introduction demonstrated the need for a supplementary form of understanding and showed that a mathematical formalization as such is necessary but not sufficient for a full understanding even of mechanical systems. Newton's "ether" falls short of indicating how ether becomes an intrinsic part of two formally related bodies such as planets, how it carries mechanical force from one planet to the other, and how it produces an attraction between two masses and not repulsion, as in the billiard-ball model of mechanical interaction. Therefore, we are dealing here not with real conceptual continuity but with a type of a hypothetical substitute of connectedness that became prevalent in the mechanistic interpretations of societies as discussed in the later sections of this chapter. In contrast, real conceptual continuity was demonstrated not only in the billiard-ball model of direct force transmission but also in the later studies of physical model systems described in Chapter 3. The uncertainty about the nature of the interaction between two masses over a distance led to reassessments of the Newtonian interpretation of gravitational attraction, as discussed in Chapter 3. The idea of the ether, as then understood, was definitely abandoned by the Michelson-Morley experiment and by Einstein's special theory of relativity.

Newton's system of thought was an advance of such magnitude that it dominated scientific progress for more than a century (Cohen 1980; Gabbey 1990; Porter 1990; Schaffer 1990). That such abstract thinking could elicit wide interest became evident when Newton's *Principia* was printed in five different editions between 1728 and 1756 and was translated into several languages. The *Principia* presented a broad readership with the close similarity of celestial and terrestrial processes and implied the universality of the basic laws of physics. What had so far been remote, mysterious, and knowable only by God was suddenly placed in the hands of the common individual.

In the words of T. M. Porter, "The Newtonian world was a well ordered machine. His astronomy showed how natural philosophy could be used to cut through the appearance of complexity and reveal the simple reality underneath. It was a model both for understanding society and for reforming it" (1990, p. 1025).[4]

In a broad sense, the establishment of a matter-force-motion relationship became the dominant trend in other areas of physics, apart from separate studies of gravitation. Thus, Newton himself thought of light as consisting of particles that, in their motions, followed his laws. He saw the possibility to "derive the rest of the phenomena of Nature by the same kind of reasoning from mechanical principles (as in the case of gravitation), for I am induced by many reasons to suspect that they may all depend upon certain forces by which the particles of bodies . . . are either mutually impelled towards one another . . . or are repelled and recede from one another" (quoted in Smith 1990). The overall impact of Newtonian mechanics is a radical simplification of reality, cogently stated by Blanpied (1969).

> The Newtonian concept of motion and interaction as the key to the understanding of the nature of things in the physical universe is the heart of that classical physics which grew out of Newton's Principia. But the ultimate logic of the Newtonian approach seems to lead to a totally mechanical model of the universe. If the nature of all things can be ultimately described in terms of the interactions of a few primary attributes, and if these attributes and interactions can be known to arbitrary precision, then all physical phenomena, and therefore, all biological and, perhaps, all psychological phenomena as well, should in principle, be completely describable in terms of matter and motion. Knowledge of the position and the velocity vector of every body in the universe at one particular time would permit knowledge of the entire past history and the entire future fate of each body, and therefore, the entire course of the universe itself. The universe becomes a gigantic clock that, once wound, proceeds along one inevitable and predetermined course (p. 186).

The rigorous form of the preceding quotation defining the ideal nature of Newtonian mechanics was chosen here because it indicates the origin of an absolute determinism in an increasingly abstract, simplified, conceptually unified form of reality. Newtonian mechanics became a signpost pointing to a rational understanding of all of the universe including man himself. Essentially, it was the abstraction from complexity and the substitution of the simplifications of the ideal systems of physics that inspired attempts to discover corresponding simplifications of sociopolitical reality. It became apparent that the scientists of that time, the natural philosophers, agreed on the adequacy of the representation of at least one aspect of physical reality. This suggested that a similar agreement could be reached by a scientific study of sociopolitical problems. The apparent conceptual and practical success of mechanics engendered the expectation, perhaps even a new and different faith, that ultimately rational analysis would not only lead to an understanding of both the inanimate and animate world but also yield an understanding of human existence, and even of God himself (Gabbey 1990; Schaffer 1990; Tamny 1990). The Age of Reason, the seventeenth through the eighteenth centuries, is thus considered here as a period in which the reconstruction of reality may be distinguished by attempts to adopt the concepts of mechanistic physics to the solution of social problems (Cassirer 1955; Berlin 1956; Gay 1969; Brinton 1972). But can human existence—individual or collective—actually be represented by a mechanistic or any other abstract, general simplification? As indicated in the following paragraphs, the Enlightenment considered both the affirmation and the negation of this question. In later periods, mechanistic simplifications remained confined to physical systems,while sociopolitical organization was represented by higher levels of complexity.

THE MECHANISTIC INTERPRETATION OF HUMAN EXISTENCE
AS A FORM OF HYPOTHETICAL RELATIONSHIP

Hobbes's Interpretation of Social Reality

An extensive use of mechanistic concepts in constructing a sociopolitical reality was initiated by Thomas Hobbes (1588–1679). He was very well educated in a strict academic curriculum and subsequently became a tutor and general advisor to noble and royal families. His life coincided with a tumultuous period in English history, and his career in England was interrupted by years in exile in France.[5] Accordingly, Hobbes's thought had two sources. From his early interest in Thucydides's *History of the Peloponnesian Wars* (Hobbes translated this work) and a later account of the English civil war in his *Behemoth,* he gained an awareness of the historical process. Hobbes's involvement in Euclidian geometry, Galileo's analysis of terrestrial motion, Kepler's mathematical representation of the planetary paths, and Harvey's explorations of the anatomy and physiology of the human body gave him a sense for deductive thinking and scientific methodology.[6] Galileo's and Kepler's work became the basis for

Hobbes's thinking about the problem of the relationship of the individual to society. It was a problem that increasingly began to occupy his mind (Peters 1972; Strauss 1973; Dietz 1990).

In a series of treatises, Hobbes dealt with the use of mechanistic concepts in the construction of sociopolitical reality: *The Elements of Law* in two parts, *Human Nature* and *De Corpore,* and in an appendix titled *Short Tract on First Principles. De Corpore* was an elaboration of an earlier rendition published under the title *De Cive* (The Citizen). During the later part of his life Hobbes published his main sociopolitical treatises, *Behemoth* and *Leviathan.* He envisaged these works as the beginning of an inclusive system of philosophy to be called *The Elements of Philosophy* that, starting from the principles of natural science (physics), would lead to an understanding of human nature (psychology), ultimately providing a guide to the principles of civil associations and governance. Whether Hobbes actually achieved such a synthesis is still debated (Watkins 1965; Herbert 1989; Kraynak 1990).[7] Also an open question is whether Hobbes used "scientific" thinking in the sense in which it was introduced by contemporary scientists, such as Galileo and Harvey, or in a sense that was essentially metaphorical.[8] In this context it will be examined how simplifying abstraction or complex concreteness enter into Hobbes's perception of reality.

Despite his exposure to experimental science, Hobbes did not free himself entirely from the influence of Descartes. Descartes followed an approach to the understanding of reality that consists of introspective abstraction, in contrast to the approaches of Galileo and Harvey who start with observations of external objects as the basis for logical reasoning (deductive thinking). Turning inward instead, Hobbes tried to recognize in his mind the basic categories of nature.

According to Hobbes's scheme of abstraction, in order to understand phenomena of any complexity—physical, biological, psychological, or social—one must resolve them first into their component parts, or, to use Aristotelian terms, their ultimate causes (Watkins 1965, pp. 52, 66–67). Hobbes calls this process resolution. Resolution is to be followed mentally by reconstituting the original phenomenon from its parts, or ultimate causes. Hobbes's resolutive-compositive procedure may be analogous to analytical and synthetic thinking as it is practiced in modern science. For example, in geometry a right-angled triangle is resolved into the three sides and the enclosed angles. The triangle is then reconstituted deductively by the relationship of the Pythagorean theorem. In physics, the trajectory of a projectile or the path of a spacecraft is the mathematically defined compositive (synthetic) result of the two resolutively (analytically) separated forces, gravity and the propelling force. In physiology

(Harvey), the circulation of the blood is the compositive result of the resolutively (analytically) recognized activities of the circulatory components: the
heart, the arteries, the veins. To Hobbes, the method of resolution and composition of phenomena leads to an understanding of the relatedness of a range of
phenomena from physics to sociopolitics. It is a framework "of consequences"
for the scientific representation of all nature as a system of hypothetical correlations.[9] In particular, Hobbes surmised that the resolution of social wholes into
their separate constituent parts would reveal the general principles that permit
the total social system to be reconstituted by logical reasoning. This type of
analysis would produce more reliable inferences than would the inductive
reasoning advocated for science by Bacon or for sociopolitical systems by
Machiavelli (Watkins 1965, p. 67).

In an initial approach to reality Hobbes was concerned with spatial extension
and its elements, such as points and particles, and with the relationship of those
elements to measurable dimensions (Herbert 1989, p. 28). This approach also
includes consideration of universal, self-evident, and fundamental concepts
such as space, time, body, and cause and effect. Later on, Hobbes developed a
more dynamic conception, in which motion becomes the basic principle of all
forms of reality. To indicate the generality of this principle of motion, he
introduced the concept of *conatus,* or endeavour (Herbert 1989, p. 46).[10] The
concept of conatus stands for a general form of dynamics that can be an element
of physical, psychological, or sociopolitical systems (Kraynak 1990, pp. 129–
133). In this sense Hobbes considers the dynamics of physical attraction and
repulsion to be analogous to psychological manifestations of appetite, pleasure,
and love and their opposites aversion, pain, and fear.

Thus, conatus becomes a unifying, general, and therefore simplifying abstraction in Hobbes's interpretation of reality. One could compare it to the
modern concept of energy, with its different manifestations in physical, biological, psychological, and sociopolitical processes. Additionally, conatus suggests
that a hypothetical correlation between the motions of the smallest elementary
particles, for example the carriers of light (Herbert 1989, p. 38),[11] and the
dynamics of the human passions plays a fundamental role in Hobbes's view of
the social process. He regards the main passions as the fear of being deprived
of the means for sustenance and the fear of violent death. These are the passions
of the common man who yearns all his life for security. On the other hand,
there is the drive by certain individuals to be acknowledged by society for their
real or imagined superiority. The latter Hobbes classifies as "vainglory." The
passions are expressed as opinions in everyday speech, as well as in more formal

discourse and political debates. According to Hobbes, opinions are statements only of apparent truths that are actually not substantiated. In particular, this is the case for opinions that are derived from extranatural sources, such as theological and authoritarian doctrines.

The rejection of doctrinaire opinions as a type of objective knowledge plays a great role in Hobbes's approach to sociopolitical reality. In Hobbes's view, opinions are presented as truth by authoritarian individuals to achieve dominance or to incite rebellions, and opinions are communicated as truths without actual evidence for their truthfulness. The effectiveness of a doctrine does not depend so much on its truth content as on how persuasively it is delivered (Kraynak 1990, pp. 69–94). Opinions are conveyed by deceptive or meaningless words instead of by clear reasoning. The doctrinaire speaker will arouse the passions of his audience and blur rationality. Man's emancipation from the oppression of doctrinaire opinions was one of Hobbes's main aims in his attack on pre-Enlightenment society. He asserts that in the past (before the Enlightenment), societies were dominated by individuals who ruled by imposing opinions, cloaked as doctrinaire truths, on a subservient population. The doctrines were absolutist and uncompromisingly exclusive, as in some monarchies or certain forms of fundamentalism. They were misleading, but they appealed to common people as simplifications of reality.

The passion-opinion relationship played a decisive role in the development of Hobbes's ideas about governance. Thus in Hobbes's system, the passion of fear generates opinions concerning the nature of the "powers invisible" (after Kraynak 1990, p. 89). Among these opinions are the pronouncements by the clergy that mitigate the fear of death by the promise of some form of existence in the hereafter. Similarly, secular authorities imply in their rhetorical opinion a greater security for the worldly existence of the ordinary citizen (Hobbes 1975, chap. 12). Hobbes exposes the whole hierarchy of fear-relieving authorities: priests, prophets, political visionaries, and "experts" who impress the fearful individual with the pretense of their special abilities to mitigate fears by simplistic doctrines. Those who exert this relieving influence often attain their status through vainglory, the passion to excel above others and to be acknowledged by the populace. The drive to vainglory can arise from a need to compensate for a feeling of inferiority or from a feeling of superiority (Kraynak 1990, p. 90). Vainglory rests on flattery by others, but there is an alternative glory that rests on real attainment called magnanimity. Vainglory requires recognition by others; real glory does not require recognition by others but is satisfied by the attainment of excellence.

Up to this point we have been concerned with Hobbes's representation of physical reality and some aspects of man's psychological nature. The questions now arise of how significant all this is for understanding man as a social being; to what extent does it provide a basis for elaborating Hobbes's sociopolitical ideas; and, is there any convergence between Hobbes's simplifying generalizations of mechanics and the complex specificity of historical, sociopolitical events?

Hobbes's understanding of human social nature is based on two abstract representations of human existence: first, the state of man as a socially independent individual (the "State of Nature"), and, second, the state of man in Hobbes's socially organized commonwealth. It was necessary for Hobbes to introduce the concept of an abstract State of Nature to strip man of the social restraints imposed by authoritarian opinion derived from extranatural or suprahuman sources. The State of Nature is understood by a process of abstraction that is a form of resolution, in the sense of Hobbes's scientific method. In the State of Nature, a person's only aim is the maintenance of the individual self, as he or she competes with all others for the means of subsistence. A person's actions are dictated by the two motivating passions mentioned earlier: fear of deprivation and brutal subjugation by others on one hand, and the "glory" of superiority of those who excel and establish themselves in a dominating position on the other hand. We have a right to acquire what we need for our subsistence. This is a natural right or "Right of Nature."

In this State of Nature no rules limit natural rights; the weaker individual is not protected, nor is the stronger individual restricted in his or her acquisitive drive (Hobbes 1975, chaps. 13 and 14, pp. 66–68). The natural right of each individual is only limited by the natural rights of all others jointly. Historically, Hobbes regarded the State of Nature as a condition of barbarism that preceded the emergence of some social structure (Kraynak 1990, p. 143). Also, a State of Nature is resumed when the structure of society breaks down, as, for example, in civil wars. An individual's behavior in the State of Nature is the expression of a desire for security, material gain, and social superiority. The concepts of justice and law are introduced only after civilization reaches a certain level. Hobbes postulates that the State of Nature is incompatible with human existence. In the State of Nature, individuals are equal; without the introduction of some restraining conditions, they will all strive for the possession of the same goods, giving rise to quarrels and violence. Human existence is only possible by the introduction of restraining rules or laws. These laws must be conceived by reasoning and must not be prescribed by authoritarian doctrines or opinions.

As the first law of reason, as Hobbes calls it, "Every man ought to endeavour peace, as farre as he has hopes of obtaining it." From this fundamental Law of Nature, by which men are commanded to endeavor peace, is derived the second law: "That a man be willing, when others are so too, as farre-forth as for Peace, and defense of himself he shall think it necessary, to lay down this right to all things; and be contented with so much liberty against other men, as he would allow other men against himself. This is the Law of the Gospel: Whatsoever you require that others should do to you, that do ye to them" (Hobbes 1975, p. 67).

Individuals divest themselves of part of their liberty in accepting a condition that limits their rights. The transfer of natural right collectively, by mutual agreement between individuals of a social group, is a covenant.[12]

The covenant is the first step in the transformation of the State of Nature into a Civil State, or the State of a Commonwealth. In the next step, the covenant must be enforced by a common power that can defend the group-in-covenant against a common enemy and can keep individuals from transgressing the agreements reached in establishing the covenant.

Transition into a State of Commonwealth is achieved by a social contract, agreed to by consensus of the common citizen and implemented by a sovereign whose absolute powers are not derived from extranatural authority. The sovereign is chosen by the common citizens. In this role he is invested with absolute legislative power, and the laws he decrees cannot be contested by the citizens of the Commonwealth. The actual implementation of the laws and other administrative duties are performed by officials appointed by the sovereign.

In extending the basic principles of mechanical motion to an interpretation of human existence, Hobbes creates a dichotomy between two abstractions: (1) motion converted into the passions of fear and boldness gives rise to individuals in their unrestrained "State of Nature"; (2) the apparent incompatibility of unrestrained individualism with the maintenance of social organization gives rise to the quasi-authoritarian doctrine of the civil state, the State of Commonwealth, and the concept of the absolute sovereign. Starting with a simple view of mechanical science, Hobbes here applied the same rationalization used in the representation of ideal systems of physics to define high-level complexity in human society.

This type of simplifying abstraction has attracted most of the attention to Hobbes's work and has generated an extensive literature dealing with its shortcomings (Peters 1972, 4:39). However, Hobbes considers both the simplifying abstractions of his resolutive-compositive procedure and the specific instances subsumed under those abstractions. He starts with a model system and uses it to predict the motion of a body in a specific natural setting. Not only the basic mechanical principles but all specific factors affecting that motion (such as friction and interactions with other bodies) must be known (Watkins 1965, p. 38). Therefore, according to Hobbes it is difficult or even impossible to know all the conditions that impinge on a body moving in a nonabstract environment. Since the identification of all factors determining a social process is even less likely than the identification of all factors in a physical system, Hobbes rejects Machiavelli's idea of a predictable social process. The distinction between abstract simplified systems and concrete complex events corresponds to the ideal and nonideal systems discussed in Chapters 3 and 4. It is noteworthy that Hobbes not only gives a general outline of his scheme of governance but that he includes a large body of detailed directions for the implementation of his ideas, the actual organization of its administration; and also devotes much effort toward clarifying the compatibility of his ideas about society with the religious traditions of his day.

Restating its main points, Hobbes's system of thought introduces a radical break in the development of the representation of reality by rejecting the existing authoritarian dogmas promulgated by church and state, scripture and tradition. For Hobbes, such dogmas express arbitrary opinions generated by passions, mainly fear and vanity. In the form of dogmas, these opinions become the mental medications dispensed by those who seek power or some other form of social superiority.

Hobbes replaces authoritarian opinion with concepts supposedly derived from rational, mechanistic physics. He uses a resolutive-compositive method to define the elements that constitute the observed object or process and that are common to apparently diverse entities. The main element pervading all forms of reality is motion. Motion is not only the basis of all physical processes, including interactions between bodies at a distance,[13] but also also gives rise to states of mind, expressed in the passions, in particular those of fear and vanity which in turn lead to social dominance or social submission. To mitigate the social effects of uncontrolled passions, Hobbes proposes a covenant under which the citizens of a state delegate part of their individual rights and liberties to an appointed individual who becomes, as sovereign, the legislative warden for the maintenance of the Commonwealth in what remains an individual-state dichotomy. In this interpretation of reality motion becomes the most universal abstract principle for the simplification of reality. In this sense, one could regard Hobbes's representation of sociopolitical reality as the equivalent of the Copernican revolution in celestial mechanics.

The extension of Hobbes's system of thought from the mechanistic motion of subperceptible particles to the governance of human societies makes his work remarkable in his time or any other. Even though conceptual continuity does not exist, Hobbes himself apparently perceived reality in the form of unifying, and hence simplifying, abstractions. Hobbes's system is more inclusive—with mechanics, psychology, and sociology as components—than representations of reality created by other thinkers of the seventeenth and eighteenth centuries.

At the same time, the expansion of Hobbes's system of thought beyond the mechanics of the basic element of his scheme, the concept of motion and its derivative conatus, is purely speculative. Although he rejected the simplifying abstractions of traditional authoritarian, extranatural dogmatisms, he substituted instead a dogmatism that seems to be based on the objective certainty of physical mechanics. Not only is there no proof for the validity of that representation, but more disturbingly, Hobbes seems to be unaware of the essentials of scientific method, despite his attendance at Galileo's and Harvey's lectures. His concept of motion is hardly distinguishable from that of Heraclitus or Aristotle and has no more scientific validity than Aristotle's final cause or Plato's crystalline spheres. Similarly, Hobbes's assumption that fear and contention are the main characteristics of the individual in the State of Nature is quite arbitrary. His introspective approach prevents him from recognizing the disparity between his own speculation and scientific verification. Even though his scheme calls attention to the desirability of obtaining the certainty of scientific representations of reality, he fails to achieve it.

For Hobbes, giving a sociopolitical representation of human existence the appearance of a scientific basis justifies the dismissal of all other doctrines as mere subjective opinions used to gratify vainglory. Hobbes expects that the scientific certainty he seemingly introduces into sociopolitical thinking will eliminate all future discord and will establish permanent peace. His scheme is supposed to be "in accordance with natural law" (Kraynak 1990, 189). In actuality, the abstract extremes, the State of Nature and the Civil State of absolute sovereignty, represent abstract simplifications that can be regarded as precursors of more recent and familiar sociopolitical ideologies (Kraynak 1990). They are akin to any of the political "isms" of the following centuries that were continuously used to foment intellectual or military warfare between apparently irreconcilable opposites.

Hobbes's ideology eliminates divine providence as a guiding principle in understanding human nature and puts the burdens of existence entirely on the shoulders of the individual. It is understandable that in witnessing the first major scientific simplifications of nature, Hobbes, facing a new reality and unaided by extranatural forces, will be attracted by the simplest possible set of rules for coping with life's hardships and find in them a promise for the solution of all humankind's problems, a promise for the end of all calamity. The appearance of this utopia as the result of mechanistic reasoning made it a more insidious misrepresentation than those earlier utopias that had not claimed such a foundation. Indeed, the misuse of mechanistic simplifications for the representation of the complexities of human existence and its consequences is perhaps the most tragic aspect of the Enlightenment. It created, as Kraynak calls it, "the liberal and the illiberal pathologies of the Enlightenment" (Kraynak 1990, p. 203). Even so, with its apparent conceptual coherence and elaborate argumentation, Hobbes's scheme had sufficient persuasiveness at his time and for some time thereafter to offer a prescription for human existence. In this sense, it can be classified as one of the utopias that are a hallmark of the era. It found particular affirmation and further development in the writings of some of the French Philosophes. On the other hand, Hobbes's ideas were qualified by Locke and Condorcet and rejected by Rousseau, all of whom questioned the applicability of the scientific method to the resolution of human problems, and by Montesquieu and Burke, who denied that human existence could be defined by simplifying abstractions.

Hobbes's representation of sociopolitical reality has been given here in enough detail to indicate some of the problems and arguments that arise in attempts to adopt mechanistic reasoning in the representation of sociopolitical reality. The following section adds a brief survey of some of the main modifications of Hobbes's system by Locke and some of the French Philosophes. This is followed by a discussion of nonmechanistic representations of sociopolitical reality that were developed during this period.

The Mechanistic Interpretation of Reality by the French Philosophes

Hobbes's mechanistic system of thought was built on Galileo's foundations of mechanics and on Kepler's calculations of planetary paths, before the publication of Newton's *Principia*. The appearance of the *Principia* reinforced a radical form of simplifying, mechanistic thinking by a group of French Philosophes, Julien Offroy de La Mettrie (1709–1751), Claude-Adrien Helvétius (1715–1771), and Paul-Henri Thiry, Baron D'Holbach (1723–1789).

La Mettrie came to his mechanistic interpretation of human nature through his study of medicine. Having first obtained his doctor's degree from the

University of Rheims (1733), he studied further at Leiden under Hermann Boerhaave, who introduced principles of physics as a basis for medical treatment (iatromechanistic medicine). Upon completion of his studies, La Mettrie first practiced medicine and translated some of Boerhaave's writings. La Mettrie's own first work, *The Natural History of the Soul,* appeared in 1745. Through a mechanistic elaboration of Locke's thinking, La Mettrie postulated that the activities of the soul are the product of certain organs and of a *force motrice* inherent in matter. He attributed human activities usually ascribed to the soul, such as emotions, reflection and volitions, to the structure and function of the central nervous system. La Mettrie further developed this mechanistic interpretation of psychological and physiological processes in his main work, *Man a Machine* (1748). The variety of these processes is the product of sense perceptions stored, classified, and combined by the activity of cerebral structures of different complexity (Vartanian 1972, pp. 379–382).

Helvétius, coming from a highly respected family background in medicine, devoted most of his life to independent philosophical studies at his country estate in France. His main works were *De l'Esprit, or Essays on the Mind and Its Several Faculties* (1759) and *A Treatise on Man, His Intellectual Faculties and His Education* (1777). Developing Locke's system of thought to its extreme, Helvetius proposed that the entire gamut of the psychological manifestations of a personality are the result of sense perception and the elaboration of these perceptions in the human mind. He made an absolute case for an environmental determination of personality as opposed to an innate, hereditary one. Diversity of personality was due to different kinds of stimulation of the passions that are received by the individual during development. The greater the stimulation of passions, the higher the level of talents exhibited by the individual. Helvetius regarded pain and pleasure as the main controls of human behavior, thus giving rise to his hedonistic ethic: humans will act in a socially beneficial fashion if their behaviors are linked to pleasurable rewards. Practicing this hedonistic ethic was impeded by Christianity and by the inequalities imposed by the absolutistic feudal regime of the time. Despite their weaknesses, Helvetius's ideas exerted a considerable influence on later developments in socialistic and utilitarian thinking (Vartanian 1972, pp. 471–473).

Paul-Henri Thiry, Baron Holbach, was born into a German family (his original name was Paul Heinrich Dietrich), but an uncle in Paris provided him with his education, the French version of his name, his title, and an income sufficient to conduct a life devoted entirely to philosophical and literary work. Thus, he was able to associate with the great French thinkers of his time and with prominent foreign visitors: Hume, Adam Smith, Edward Gibbon, Joseph Priestley, and Benjamin Franklin. His initial writings were contributions of articles to the French *Encyclopedie* on aspects of chemistry, mineralogy, and geology. Geological evidence he found for the development of the earth, negating the doctrine of creation, promoted his atheist and anticlerical articles. Bringing together the ideas of Helvetius, Diderot, La Mettrie and others, he became the most outspoken exponent of atheistic materialism in his *Christianity Unveiled; Being an Examination of the Principles and Effects of the Christian Religion* (1795).

According to Holbach's mechanistic representation, Nature is fully defined by matter and motion. In an eternal universe all change is due to conversions of matter or energy, with corresponding changes in combination of the elementary constituents of matter. All changes, including man's actions, are fully determined from the outset. Satisfactory human existence requires the cooperation of individuals, and the main aim of sociopolitical science and governance is the understanding and adjustment of the balance between the individual as a separate unit and the individual as a part of the greater whole of society. Holbach suggests that this balance can be maintained in a constitutional monarchy by some form of parliamentary representation in which the distribution of wealth is controlled by progressive taxation. Although Holbach's representation of nature was rather simplistic, his ideas about governance show the beginning of an appreciation for the complex relationships between individuals and society. His ideas are summarized in his main work, *The System of Nature; or the Laws of the Moral and Physical World* (1795; Vartanian 1972).

These Philosophes shared a mechanistic representation of broad segments of reality, in particular that of human existence, with Hobbes. They also shared an unreserved acceptance of certainty as the main characteristic of their mechanistic interpretation of human reality. Although they followed Locke's premise of sense perception as the basis of individuality, they disagreed with his rejection of mechanistic certainty and his substitution of a probabilistic understanding of human nature. But Locke's probabilistic approach was taken up and further developed by Marie Jean Antoine Nicolas de Caritat, Marquis de Condorcet, one of the youngest of the French Philosophes (1743–1794; see Baker 1972, 1989). In a sense, Condorcet's career reflects the tragedy of the Age of Reason in its entirety. After completing a Jesuit college curriculum, Condorcet devoted himself to mathematical studies, primarily by participating in the elaboration of the differential calculus, an active pursuit of the French mathematicians of the time. Jean de Rond d'Alembert, physicist and mathematician who had been working with Newton's concepts of motion, found Condorcet's work of interest and promoted his membership in the prestigious French Academy of Sciences. Condorcet was admitted to membership in 1769, and he became the Academy's permanent secretary in 1776 (Baker 1972, 2:182). D'Alembert also introduced Condorcet into the salons of Paris, the social centers of French intellectual life.

As a member of the Academy and through his connections with important political figures such as Turgot, Condorcet began to explore the possibility of developing a coherent system of thought on social and political problems. At that time probabilistic thinking was beginning to be used in defining various societal characteristics, such as life expectancy and birthrate. Condorcet set out to apply probabilistic thinking to an analysis of majority decisions made by political assemblies or under juridical procedures. He presented his approach in *An Essay on the Application of Analysis to the Probability of Majority Decisions* (1785) and in a later unfinished treatise, *General View of the Science Comprising the Mathematical Treatment of the Moral and Political Sciences* (1793). Through his probabilistic treatment of sociopolitical systems he hoped to achieve an increased reliability in the evaluation of social problems, similar to that achieved by mechanistic thinking in the understanding of physical reality. Condorcet thought he could introduce a compelling method of reasoning into the everyday affairs of politics, but he encountered a basic dilemma, indicated by the following quotation:

> Everyone regards himself as judge; no one imagines that a science employing the terminology of everyday language needs to be learned; the social right to have an opinion on social matters is confused with the right to pronounce on the truth of a proposition, which enlightenment alone can give. One wants to judge; and one is mistaken.

As Baker (1975) points out:

> Here Condorcet raised one of the cardinal issues in his conception of social science. For Condorcet, the professional academician, the goal of science was to transform societal choice into the rational decision-making of the idealized republic of science in which individual opinions were not counted but weighed. For Condorcet, the theorist of liberal democracy, the right of each citizen to an equal voice in social decision-making came to be one of the natural rights of man. This tension between scientific elitism and democratic liberalism—between rational choice and popular will—lay at the heart of the philosophes' thinking (p. 193)

Condorcet attempted to resolve this dilemma by submitting concrete, specific proposals for governance to the legislature during the revolutionary period (see Chapter 6), after joining the conservative party of the Girondins, thus confronting the radical Jacobins in the National Assembly. With the takeover of political power by the Jacobins, Condorcet's cause of governance by reasoned compromise was lost, and Condorcet went into hiding for several months to escape the Jacobin terror. On returning to public life he was arrested and jailed. Two days after his arrest he died from exhaustion, or suicide.

Condorcet's life is symbolic of the fate of the Age of Reason. As described in Chapter 6, an initial reasoned compromise began in the French Revolution. It was soon superseded by the action of willful individuals in their striving for power. The Philosophes' faith in the certainty of rational mechanistic thinking was replaced by the people's commitment to nationality, inspired by the charisma of Danton, Robespierre, and, last but not least, Napoleon. While the search for the unification of physical reality reached its climax and the reasoned approach to human affairs lingered on in academia, Western civilization spawned nationalism and Marxist communism as ideological forms of sociopolitical reality, as will be described in Chapter 6.

Even during the eighteenth century, however, mechanistic thinking did not remain the sole conceptual basis for the interpretation of sociopolitical reality. Locke rejected the mechanistic determination and substituted a probabilistic approach toward understanding reality. Rejecting the possibility of a strictly rational understanding of human existence, Rousseau intuitively took a step toward sociopolitical complexity by giving the individual and the state a nearly balanced and flexible relationship. At about the same time, more complex alternatives to the simplifying mechanistic representation of reality were introduced by Montesquieu and Burke.

LOCKE'S EMPIRICAL INDIVIDUAL AND THE SHIFT TO THE PROBABILISTIC SIMPLIFICATION OF REALITY

Hobbes's primary concern was with the state as an organization of individuals, rather than with individuals themselves; Locke's primary object of social reality was the individual, with the state as a secondary superstructure. Hobbes did not question the certainty of mechanistic reasoning in his interpretation of social reality. Locke limited mechanistic reasoning to the abstractions of mathematics and geometry and introduced a probabilistic interpretation of nonmathematical reality. One might surmise that Locke's interest in the nature of the individual was more compatible with his study of medicine and his collaboration with the noted medical authority Thomas Sydenham, rather than with the abstractions he had dealt with in his study of the Greek and Latin classics at Oxford, which he disdained.

In his later career Locke became associated with the Earl of Shaftesbury, and for a short time held a position as secretary of one of the governmental agencies. He lived for several years in Paris (1675–1679) and met the more outstanding French thinkers. His association with Shaftesbury, who led the parliamentary opposition to the Stuarts, put him into a precarious political position, and both he and Shaftesbury took refuge in Holland. Both returned eventually to England in 1689 when William of Orange assumed the English throne with Locke as his advisor (J. G. Clapp 1972).

Locke's primary interest was the representation of reality in the human mind. In accordance with a mechanistic simplification, and as an example of a hypothetical relationship, the mind was to be thought of "as if it were a box containing mental equivalents of Newtonian particles." These were called "ideas" and were conceived as distinct and separate entities that are "simple," that is, possessing no parts into which they can be split (Berlin 1956, p. 18).

"Ideas are the objects of understanding when a man thinks . . . whatever is meant by phantasm, notion, species, or whatever it is, which the mind can be employed about thinking" (Berlin 1956, p. 36; Clapp 1972, 4:490). Included in the category of ideas are simple qualities external to our mind, for example, whiteness, hardness, coldness, or even more complex entities such as chair, horse, or rose. Locke contended that ideas are perceived through the senses. Simple ideas can be transformed by association into new, more complex ideas: the simple ideas of whiteness, coldness, and roundness can be put together to yield the more complex idea of a snowball. Another set of ideas represents the operations of the mind; among these are thinking, believing, doubting, and reasoning. These ideas are the result not only of sensation but also of reflection. All ideas are ultimately derived from experience. Since each individual starts his existence without ideas, and since each individual has different experiences, divergent sets of ideas are established in different minds; thus, individuals differ from each other. All ideas are the direct result first of sense perception and then of subsequent reflection.

There are no innate ideas, however, according to Locke. Based on rather weak argumentation, Locke made his denial of innate ideas the basis of his representation of reality. If innate ideas exist, innate ideas should be common to all people. Since such universally recognized ideas had not been found, Locke held the existence of innate ideas to be unlikely. On the other hand, ideas supposedly held in common, such as the basic principles of logical arguments, did not appear to be inherent in the minds of children or uneducated adults.

Of particular relevance here is Locke's departure from the mechanistic thinking of his time in limiting certainty of understanding to mathematical and geometric relationships. With certainty we can expect that in all right triangles the square of the hypotenuse will be equal to the sum of the squares of the other two sides because this relationship necessarily follows from the logic of the Pythagorean theorem. Locke points to the clockmaker's certainty in his understanding of a clock mechanism as a nonmathematical example. In this case the certainty is not an abstract mathematical relationship but the identity of known direct mechanical interactions between the parts of the clock that provides a clear example of conceptual continuity, and also the basis of certainty. In contrast, although it was observed time and again that gold dissolves in acids such as *aqua regia,* and silver in *aqua fortis,* in Locke's time it remained unknown how this process took place and of what the mechanism of this transformation consisted. There was no logical necessity to expect its recurrence with certainty. The real conceptual continuity between the acid solvents and the dissolved metals was lacking, and in its absence knowledge was a matter of probability, not of logical certainty. Similarly, the postulated interactions between the planets and the sun by gravitational force lacked the logical necessity of the Pythagorean theorem or even the mechanical necessity of the clockwork and, in Locke's view, had to be regarded as relationships that only existed with high probability, not with certainty. As discussed in the next chapter, only the atomic models of quantum mechanics established conceptual continuity in thinking about the interaction of acid and metal (and chemical reactions in general), but the full understanding of the interactions of masses over a distance by a gravitational wave mechanism is still lacking.

Locke's distinction between certain knowledge and probable knowledge inaugurated an alternative way of thinking that became especially useful for considering sociological problems. In his *Two Treatises of Government* Locke put individuals into a social context. As many others before and after him, he again postulated a fictitious initial State of Nature, "a state of perfect freedom [in which individuals may] order their actions, and dispose of their possessions, and persons as they think fit, within the bounds of the Law of Nature, without asking leave or depending upon the Will of any other man" (Clapp 1972, pp. 498–499). Individuals live peacefully in this state without infringing on the rights of others; those infringements are prevented by a social contract based on common consent between free individuals, without the ruler-ruled relationship.

"The aim of the contract is to preserve the lives, freedom, and property of all as they belong to each under natural law" (Clapp 1972, p. 499). Property is the product of each individual's labors, but no one is allowed to amass property at the expense of others. The establishment of laws regulating the use, distribution, and transference of property is a primary aim of social organization and supersedes the State of Nature. The social contract is now one of a new type, one in which individuals give up not all their rights, just the legislative and executive powers they originally had under the Law of Nature. Sovereignty ultimately rests with the majority of people who "establish the legislative, executive and judicial powers" (Clapp 1972, p. 499). Under such conditions governance can take the form of monarchy (except absolute monarchy), oligarchy, or democracy. In view of the political events of his lifetime, Locke admitted the right of the people to rebel against any form of usurpation of absolute power. Thus, Locke's *Two Treatises* was a step toward enhancing the role of the individual citizen in society in a relatively simple representation of sociopolitical reality. Locke's thoughts were of great relevance for the development of democratic governance (see Chapter 6), and in this sense they represent an initial step in the introduction of conceptual continuity into sociopolitical reality.

THE SEARCH FOR CONCEPTUAL CONTINUITY BETWEEN INDIVIDUAL AND SOCIETY: ROUSSEAU'S INTRODUCTION TO SOCIAL COMPLEXITY

Jean-Jacques Rousseau (1712–1778) was a person of many and great talents. Early on, Rousseau began to follow a checkered course (see J. Conaway Bondanella 1988; Grimsley 1972). He was self-taught in many disciplines, including mathematics, science, astronomy, music, and literature, and he eventually became acquainted with the political philosophies of Hobbes and Locke. The multidisciplinary development of Rousseau's personality found expression in his creative work. He contributed two novels of social criticism, *Emile* and *The New Heloise;* a musical entitled *The Village Soothsayer;* a *Dictionary of Music;* and several articles for the *Encyclopedie.* He described the vagaries of his life with unusual candor in his autobiographical *Confessions.* Rousseau's general sociopolitical writings, embodied in four *Discourses* and additional fragments dealing with specific instances of governance in Corsica and Poland, are relevant here (Grimsley 1972; Ritter and Bondanella 1988; Graubard, 1978).

The introductory remarks to his works suggest the multiple origins and different directions of Rousseau's creativity. He opposed the mechanistic reductionism (Hobbes) that, to him, deadened the imaginative renewal of man's existence. At the same time, his flights of imagination had to remain compatible with the dicta of reason. In the *Discourses,* Rousseau turns from the mechanistic simplifications of the sociopolitical systems examined so far to a somewhat higher level of complexity by emphasizing the duality of man's existence as a distinctive self and as a subordinate entity in the greater whole of the political state.[14] Here Rousseau takes a step from the single plane of mechanistic reasoning to the double-layered reality of the individual-state relationship. Man no longer embodies the uniform, subservient dignity allowed him during the feudal Middle Ages; man is not an unspecific component of a general and Newtonian mechanism as envisioned by Hobbes or the Philosophes; nor does he have the unqualified, independent individuality attributed to him by Locke (Featherstone 1978).

How does Rousseau come to terms with this ambivalence? His point of departure is his dissatisfaction with contemporary society. He exposes man's social condition in his *Discourse on the Sciences and Arts* as the unrestrained competition between individuals for money, luxury, and power, with its accompanying corruption and frustrating inequality. Rousseau regarded this inequality as the main source of an intolerable societal condition. Understanding the origins of this inequality was for Rousseau a key to the development of a better form of social organization. Rousseau begins by asking first how man existed before the rise of inequality and then proceeds to the fundamental problem of how inequality can be eliminated without suppressing individuality. He answers the first question in his *Discourse on the Origin and Foundation of Inequality.* In this discourse he invokes idealized early forms of society, in which he includes prehistoric man, Sparta, and the first (idealized) phase of Roman history. His idealizations are, of course, completely lacking in historical reality. In these almost legendary societies human beings lived, he suggests, in harmony: in a "union of hearts" (Starobinski 1988, p. 221) in full communication with each other. It is this social condition that Rousseau designates as the "State of Nature," a fundamentally different view from that of Hobbes.

The main characteristics of human beings in Rousseau's State of Nature are the satisfaction of their basic needs by the resources found in their immediate environment and the absence of desires beyond these basic needs.[15] Rousseau assumes that people are at peace in the State of Nature since any incentive for strife, and especially the violence of war, would be absent. At the same time, the need for social interactions seems minimal. Humankind may exist without structured language; problems of morality and of social competitiveness are absent; it is a state of happy innocence. This is in stark contrast to Hobbes's conception of the State of Nature in which the fear of losing the means of subsistence leads to fierce competitiveness and violence. Rousseau leaves it an open question whether at any time man existed in such utopian harmony. For him the hypothetical concept of the State of Nature established a basis for comparing societies, including his own, at different stages of advancement.

In Durkheim's (1960) interpretation of his work, Rousseau did not think that man was an innately social being. Society was generated by external circumstances. Thus, social interaction may have arisen through the necessity of sharing resources such as food and water during times of scarcity, and through the need to cooperate in hunting and in domesticating animals. The initial benefits of cooperation soon led to competitive strife, however, and this led to inequality among men. Corruption set in with the advancement of civilization. Individuals no longer lived with each other but against each other. The temptation to exert power and to appear socially superior are in conflict with the social essence of harmonious coexistence. A loss of the feeling of unity with others and within oneself leads to self-alienation and engenders historical pessimism. The laws and conventions that regulate social interaction and strife transform a loose association of individuals into an entity with its own, different characteristics. It is a whole that has become distinct from the sum of its unrelated parts.

Rousseau ascribed inequality, the main source of conflict in advanced society, to unjustified exploitation. He expanded on this thesis in his *The Citizen: Discourse on Political Economy* published in 1755. Seven years later Rousseau attempted to give in his *On the Social Contract* (pub. 1762) an answer to the question of what kind of government would resolve the conflict between the individual and society. He asserted that at some point in the historical development of society, the State of Nature becomes insufficient for man's survival. A social contract arises when existence in nature as an independent individual becomes too difficult. Individuals have to band together, they must unite, to generate the forces needed to preserve the species.

In contrast to Hobbes, in Rousseau's state power is not delegated to a single individual (the sovereign), but rests with a legislative assembly. The laws set forth by this body are administered by an executive branch of professional magistrates, separated from the legislative process. In appropriating the seigniorial right to make and repeal the laws regulating the existence of the association itself, the legislative assembly becomes the body politic, and assumes the role of the sovereign. Every member of the association has the right to participate in the legislative process provided he assumes the duties and responsibilities as a subject of the self-inaugurated state.

In Rousseau's state the only legitimate source of the law is the sovereignty of the people, defined as the "General Will."[16] The individual becomes a citizen by participating in the declaration of the General Will. It remains unresolved, however, whether a minority can represent the General Will (Starobinski 1988, p. 228). A passage from the *Social Contract* is included here at some length because it is crucial for the further evaluation of the relationship between the individual self and the communal self.

To become an associate (citizen and/or subject) requires:

> The total alienation of each associate with all his rights to the whole community. For, in the first place, since each person gives himself entirely the condition is equal for all, and since the condition is equal for all, no one has an interest in making it burdensome for the other.
>
> Furthermore, since the alienation is made without reservation, the union is as perfect as it can be, and no associate has anything more to claim. For, if some rights were left to private individuals, and there were no common superior who could decide between them and the public, each person, being in some respect his own judge, would soon claim to be so in every instance; the State of Nature would subsist and the association would necessarily become tyrannical or ineffective.
>
> Finally, each person, in giving himself to all, gives himself to no one, and as there are no associates over whom he does not acquire the same right as he concedes to them over himself, he gains the equivalent of all that he loses and more force to preserve what he has.
>
> If, then, we eliminate whatever is not essential to the social pact, we shall find that it can be reduced to the following terms: Each of us puts his person and his power in common under the supreme control of the General Will, and, as body, we receive each member as an individual part of the whole.
>
> . . . This public person, which is thus formed by the union of all the other persons, formerly took the name of a city and now it takes that of republic or body politic, and its members call it a state when it is passive, a sovereign when it is active, and a power when comparing it to others of its kind. As for the associates, they collectively take the name of the people, and, individually they are called citizens, when they participate in the sovereign authority and subjects when they are subject to the laws of the state (*Social Contract,* In Ritter and Bondanella 1988, pp. 92–93).

It is essential to examine the implications of the term "alienation." In an extreme absolutist sense, the term indicates a complete abrogation of individuality and a uniformity of the associates through the sovereignty of the General Will. Alienation can be construed in this way out of context. In turn, this has led to Rousseau's indictment as a prophet of totalitarianism. Such a total alienation negates the duality of the individual, and all that remains is the individual as part of the whole of society.[17]

It becomes apparent in the first book of the *Social Contract* that this is an oversimplification on the part of the interpreters. Rousseau's counter-challenge is made explicit there in a few short but critical statements. Thus, it is the purpose of the *Social Contract:*

> To find a form of association that defends and protects the person and possessions of each associate with all the common strength, and by means of which each person, joining forces with all, nevertheless obeys only himself and remains as free as before (Ritter and Bondanella 1988, p. 92).
>
> What man loses by the social contract is his natural liberty and an unlimited right to everything that tempts him and to everything he can take; what he gains is civil liberty and the ownership of everything he possesses . . . another benefit which can be counted among the attainments of the civil state is moral liberty which alone makes man truly his own master, for impulsion by appetite alone is slavery and obedience to the law that one has prescribed for oneself is liberty (Ritter and Bondanella 1988, pp. 95–96).

This is further elaborated in a chapter in the *Social Contract* entitled "On the Limits of Sovereignty" in which Rousseau states:

> Beyond the public person, we have to consider the private persons who compose it, whose life and liberty are naturally independent of it. It is a question, then of clearly distinguishing the respective rights of the citizens and the sovereign, and the duties the former have to fulfill as subjects from the natural rights they should enjoy as men (Ritter and Bondanella 1988, p. 101).
>
> . . . each person alienates through the social contract only that part of his power, possessions, and liberty that will be important to the community but it also must be agreed that the sovereign alone is the judge of what is important (pp. 101–102).

"However, the particular will of the individual must not be allowed to impose itself on the community as a whole; the particular will cannot be allowed to replace the General Will. Thus, in contractual society, part of natural law remains in force" (Leigh 1988, p. 238). There is a private aspect of human existence that is not subject to the General Will.

The structure of the sovereign body politic is enhanced by Rousseau's insistence on the separation of legislative and executive functions. In his view the exercise of executive process is a highly technical matter that is not within the competence of an assembly of the people as a whole, and is to be in the hands of elected magistrates (Cobban 1934, p. 111). Rousseau also considers the judicial process to be a separate governmental function. In this respect he regards the English constitution as a model for the separation and balance of the respective governmental functions (p. 103). Yet, Rousseau is opposed to the separation of legislative function by its transfer from the citizens at large to a body of elected representatives, a feature common to other constitutional governments (p. 108).[18]

It is clear that Rousseau explicitly rejected direct popular government, which he called "democracy" according to his own definition. It is not only the illegitimate form of this government that he condemns under the name of ochlocracy, but also the legitimate form. According to Rousseau's scheme: "To be legitimate the government must not be confused with the sovereign, but must be its minister" (Ritter and Bondanella 1988, p. 107).

Further on the condemnation becomes more explicit: "It is not good for the one who makes the laws to execute them, or for the body of the people to divert its attention from general objectives and turn to particular ones. . . . taking the term in the strict sense, true democracy has never existed. It is impossible to imagine that the people would remain constantly assembled to attend to public affairs" (Ritter and Bondanella 1988, p. 125).

Finally, there is the well-known aphorism with which he concludes: "If there were a people of gods, it would govern itself democratically. So perfect a government is not suited to men" (Ritter and Bondanella 1988, p. 126).

Rousseau's outline of governance excludes political, economic, or social associations as well as the subdivisions of the official state organization. Rousseau seems to have an inherent fear that subordinate associations could exert a disproportionate influence on the governance of the state.[19] The danger of these associations soon became evident in French history when the Jacobin Club, a prototype political association, took over the government and initiated the reign of terror, the worst phase of the French Revolution.

Rousseau's extending his anti-association stand to the church seems to emphasize the totalitarian aspect of the *Social Contract*. He was concerned about a dual loyalty to church and state with the possibility that the church might introduce general rules into the fabric of a community. In Rousseau's view church and state should be unified. Since some form of religion seemed essential to him, the idea of the community as a whole should be substituted for the conventional idea of God (Cobban 1934, p. 79). Rousseau conceived of religion not as a rationalist scheme but rather as a representation of communal existence with emotional appeal. His definition of this type of civil religion became the most controversial aspect of the Social Contract:

> A purely civil profession of faith, the articles of which are for the sovereign to determine, not precisely as religious dogmas, but as sentiments of sociability, without which it is impossible to be a good citizen or a faithful subject. Without being able to obligate anyone to believe them, the sovereign can banish from the state anyone who does not believe them; it can banish him not for being impious but for being unsociable, for being incapable of sincerely loving the laws and justice, and of satisfying, sacrificing, his life, if need be, for his duty. If after having publicly acknowledged these same dogmas, someone behaves as though he does not believe them, let his punishment be death; he has committed the greatest of crimes; he has lied before the laws (Ritter and Bondanella 1988, p. 172).

In Rousseau's system of thought, the abolition of the traditional church was required because a dual loyalty, to both state and church, was inconceivable. It essentially extends the abolition of subordinate associations of any kind. The regulation by the church of such civil acts as marriage and divorce, and of moral censorship, was incompatible with the sovereignty of the state. The subordination of the church under the state had also been proposed by the Philosophes. Regulations of social conduct could not be shared by two agencies but had to remain in a single hand, for the state is to represent the community as a whole (Cobban 1934, p. 79). In place of the traditional church, Rousseau substituted an "organized system of beliefs" (p. 80) that is to be followed by all individuals who want to be citizens. Yet his horrendous proposal to impose the death penalty for disregarding the rules of the civil church were among the conspicuous aspects of his system that most decisively seemed to make him an absolutist totalitarian. However, here again, Rousseau qualifies his position: most striking are his repeated denunciations of religious intolerance. The apparent restrictiveness of the civil religion is further mitigated by his stipulation that the citizen may exercise other forms of religious confessions together with his civil religion as long as they do not interfere with the duties and responsibilities determined by the *Social Contract* (Leigh 1988, p. 241).

In considering an egalitarian distribution of wealth, Rousseau rejected outright confiscation of individual property and instead proposed a form of progressive taxation that would be "exactly proportioned to the circumstances of the individual" (Ritter and Bondanella 1988, p. 246). Here, however, Rousseau saw several difficulties in the realization of this aspect of his utopia. Essentially his utopian republic was a closed economic system. Commerce with the rest of the world would lead to disproportionate accumulation of wealth in the hands of a few individuals and hence increase the inequality of the republic's citizens.

The other barrier to economic equivalence was the increased need for a bureaucracy to administer the taxes and other methods for supervising the equalization. Whatever free time might be created by improved productivity would be devoted to civic purposes: an additional requirement for utopian citizenship. As Ignatieff points out: "Rousseau was not, as one might have expected, opposed to a society of abundance, but he insisted that such a society can never be virtuous, and its members reconciled to themselves, unless abundance is equalized. If the unchecked tendency of the invisible hand in history is to reward the propertied, the first task of modern government is to 'prevent extreme inequality of fortune'. Politics must redress the natural injustice of history" (Ignatieff 1984, p. 114).

Therefore, it is important to note that "Rousseau's redistributive politics were always cautious and prudent. He left the right to bequeath property untouched." "Nothing," he said, "is more fatal to morality and to the Republic than the continual shifting of rank and fortune among the citizens" (Ignatieff 1984, p. 114).

This could have almost been said by Burke and certainly nulls Voltaire's invectives against Rousseau (Ignatieff 1984, p. 114). Ironically, Adam Smith, the prophet of the laissez-faire, free market economy and in most respects an adversary of Rousseau, planned, in the end: "To make demands on the virtue of his utopia's participants as austere as Rousseau's. A market society could remain free and virtuous only if all its citizens were capable of stoic self-command. Without this self-command, competition would become a deluded scramble, politics a war of factions, and government a dictatorship of the rich. Smith was optimistic, but it was an optimism based on the stoic hope that the human will would prevail and each individual would retain the capacity to know the difference between what he wants and what he needs" (Ignatieff 1984, p. 124).

The *Social Contract* should perhaps be regarded as a theoretical framework, an ideal standard that has to be modified for use in practice. Overall, Rousseau's treatment of governance is not absolutist, because he suggests that every state would be governed by some form of contract best suited for its special needs. Human beings are different according to their traditions and the climates in which they live, and their systems of governance should be adjusted to those particular circumstances (Cobban 1934, p. 90). In this sense, Leigh points out (p. 140) that the *Social Contract* DOES NOT provide a blueprint which is supposed to enable us to set up uniform societies, each identical to every other, all emerging from sociological production lines like mass-produced cars. On the contrary, there can be an infinite number of societies, quite different from each other, but all legitimate according to the criteria of the *Social Contract*.

In contrast to the conventional view of Rousseau as the promoter of radical revolutionary change, his specific proposals for the governance of Poland and Corsica show that his approach to governance differed fundamentally in concrete situations. As described by Starobinski (1988, p. 230): "He did not work in the abstract in this instance. He collects documents; he surrounds himself with maps, history books, he asks for time to reflect, he wishes to know the climate, the resources, the industries, he even dreams of a trip."

All change—the abolition of serfdom, the abolition of nobility, the creation of a federation—is to occur at slow pace so that existing traditions are not disturbed.

Rousseau can be seen as trying to bridge the gaps between utopian and practical actuality and between simplifying abstract generalization and the complexity of concrete specificities. More fundamentally, what Rousseau tried to achieve is the reconciliation of the individual self (the *moi*) and the communal self (*moi commun*) as "coexistent desires, being the pathos of the human condition . . . wanting identity and community, too, and both in full measure. The yearning for a '*moi*' realized in all its potentiality and for a '*moi commun*' (despite our experience of its cruelty and distortion) is as poignant as ever" (Manuel 1978, p. 12).[20]

Rousseau's merit is his acknowledgment of the duality of human existence and the need for balancing its two aspects. Whereas his predecessors shifted the balance to one side, the communal self in the case of Hobbes, or to the individualistic self in the case of Locke, Rousseau attempted to give equal emphasis to both. It is this balanced vision of duality that defies simplification. In each society and at each level of the social hierarchy, this duality is differently and only transiently defined.

Duality as a simplifying general concept loses relevance, and the specific factors determining the complexity of the different phases of governance acquire importance, when we compare the generality of the *Social Contract* with Rousseau's blueprints for the governance of Corsica and Poland. These are close to the conservative proposals for governance made by Montesquieu and Burke, with their emphasis on specific circumstances. Rousseau makes this duality a distinctive form of reality. He left the two aspects of reality, the general and the particular, standing side by side as an open-ended proposition.

Rousseau's representation of the duality, the two "mois," was stronger than that of Hobbes or Locke and, therefore, more threatening as a form of socio-political complexity. The challenge given by the controversial nature of this duality was enhanced because Rousseau's writing included "the most brilliant passages and torch-like phrases to be found anywhere in the history of thought" (Nisbet 1988, p. 225). Rousseau's ideas were a search for the elusive conceptual continuity between individual and society that gradually evolved with the advancement of democratic governance. Perhaps more fundamentally, Rousseau's system of thought is here of such pivotal significance because it summed up the bygone achievements of the Enlightenment and generated the three main trends of thought and action that were to underlie the following century.[21]

In one of these trends, some of Rousseau's successors failed to see or dismissed the potential for an advance to a new complex reality. Instead, they found in his double-edged utopia a niche for their own respective brands of simplification in the form of revolution, nationalism, romanticism, or totalitarianism. It is ironic that Rousseau's own proposals for governance were not based on revolutions, but on conservative principles; and the revolutionary interpretation of his work came secondarily from the leaders of the revolutionary movements. In a second trend, the challenge of advancing toward more complex forms of governance was met as Montesquieu, Burke and the American Constitution secured the foundations of democratic government. A third trend consisted in the attempts to give Rousseau's visions a firmer, systematic basis. It was the move to develop, separate from mechanistic physics, an indigenous scientific program for the study of humanity. In this context Rousseau is regarded as one of the initiators or forerunners of sociology, anthropology, and psychology. It remains now to follow some of these trends in more detail. As contemporaries of Rousseau and as important representatives of eighteenth century sociopolitical thought, the approaches of Montesquieu and Burke are included.

Montesquieu, The Transition from Simplifying Utopia to Complex Political Practice

The life of Charles-Louis de Secondat, Baron de Brede et de Montesquieu (1689–1755), mostly overlaps the era of the French Philosophes. He was born in Bordeaux, and after studying law he made his career in various positions in local government. This suggests that he was grounded in the concrete detail of everyday administrative work. He also developed an interest in science and published papers in physics, physiology, and geology, and he found time to write about more general topics (Neumann 1949, p. xi). His first major work, *The Persian Letters,* was essentially a study of the nature of love and a general cultural critique. Its publication gave him an introduction to Parisian intellectual society and eventually led to his election to the French Academy. Later, he spent about ten years in travel, including a two-year stay in England. There, he became acquainted with different forms of society and in particular was able to study the English form of government. In 1731, Montesquieu returned to his hometown, La Brede, to work on his magnum opus *The Spirit of the Laws* (Montesquieu 1949). Its publication in 1748 was highly successful. However, criticisms from various sources compelled him to publish a defense of the work.

Montesquieu's thoughts were influenced by the writings of Montaigne, Descartes, Malebranche, and Machiavelli (Neumann 1949, p. xxxiii). He was a skeptical conservative, who questioned the capacity of men to generate and maintain radically new forms of society. He rejected utopias and confessed great tolerance for the varieties of human individuality and for diversity of governance. He actually suggested that the introduction of variant subgroups into established societies may lead to the correction of social abuses (Neumann 1949, p. xvi).

With this as background, we can conceive Montesquieu's work as a transition from the preceding simplifying, abstract, sociopolitical utopias to the concrete complexities of political practice, as perceived by Burke. Montesquieu was fully aware of the mechanistic basis of his contemporary science, but he did not assent to an unqualified mechanistic interpretation of society.[22] To him, both specific and general characteristics of society were of equal importance.

Up to Montesquieu's time, philosophers had postulated what societies should be like in abstract, a priori discussions. Plato, Aristotle, the doctrinaires of the Middle Ages, and several thinkers of the Enlightenment had developed general abstract models of the organization and maintenance of societies without having systematically analyzed the elements of existing societies. In contrast, Montesquieu first set out to survey the details of a great number of societies as the basis for his generalizations. He presented his observations and his thoughts about them in his major treatise *Spirit of the Laws.* He tells the reader: "This book deals with the laws, customs, and diverse practices of all the peoples of the earth. Its subject is vast, for it embraces all the institutions that prevail among human beings" (Neumann 1949, p. xviii).

He emphasizes the relevance of specific aspects of sociological reality: "I write not to censure anything established in any country whatsoever. Every nation will here find the reasons on which its maxims are founded. . . . Could I but succeed so as to afford new reasons to every man to love his prince, his country, his laws; new reasons to render him more sensible in every nation and government of the blessing he enjoys, I should think myself the happiest of mortals" (Neumann, 1949, p. lxviii).

In his explorations of society, Montesquieu does not consider the psychological aspects of man, for example, Hobbes's "passions," but turns directly to the examination of man's social state. He distinguishes, just as did several other thinkers of that period, a State of Nature and a Civil State of man. In contrast to Hobbes's scheme, and similar to Rousseau's, Montesquieu postulated that in the State of Nature, man experiences his weakness. There is a low level of aggressiveness, and peace would prevail (Neumann, xxxix). The State of Nature is again a hypothetical, general concept: it can be defined by basic laws applying to man in the State of Nature in all societies. There laws protect and control individual's rights to define and preserve life, to acquire food for sustenance, to reproduce, and to interact with other individuals in an informal fashion. Therefore, laws referring to humans in the State of Nature essentially pertain to the existence of individuals in general. There is another class of laws beyond these limited, general laws, that pertains to the organization of society as a whole. These civil laws require a consideration of the specific needs of each society. The systematic classification of the diverse types of governance as equally valid forms of social organization and as a basis for the identification of their respective needs is Montesquieu's major achievement.

On the basis of his survey of societies, Montesquieu confirmed the three traditional classes of governance: republican democracy, monarchy, and despotism. Each of these classes has its own intrinsic motivating principle: virtue, honor, and fear, respectively. Virtue is the identification of the interests of the individual with those of the state. It points to the concept of equality in line with Rousseau's idea of the General Will. According to Montesquieu, absolute equality is unobtainable, and the main aim of democracy is the elimination of excessive differences. Liberty is limited to actions that are not prohibited (Neumann 1949, p. xi). As Montesquieu observes: "Since every individual ought here to enjoy the same happiness and the same advantages they should consequently taste the same pleasures and form the same hopes which cannot be expected but from a general frugality. Concern with common welfare eliminates a source of inequality and rotation of office holders prevents acquisition of undue power" (Durkheim 1960, p. 28).

In monarchies, a greater diversity of groups of individuals is more prominent than in democracies. This includes a sharper division of trades and labor as well as of governmental functions, divided into legislative, executive, and judicial branches. There are fixed laws that cannot be modified by the monarch himself and hence limit his power. The occupational classes of citizens exert a certain restraint upon one another and their autonomy and development are limited. Under these conditions, the main concern of the common citizen is the occupational class and not the state as a whole. There are differences in the social status of individuals within each occupational class and, correspondingly, a greater motivation in striving for socially acclaimed positions. To some extent, cooperation yields to competition. Ambition leads to optimal work performance as the basis for social advancement. Striving for the highest possible status in society is designated as honor (Durkheim 1960, pp. 28–30).

Under despotism, individual or class participation in governance is eliminated and there is no virtue in individuals' giving themselves to the higher social good and no honor in striving ambitiously for the higher ranks in society. There is restraint and fear of reprisal among the ruled, except for the ruler, and a small elite appointed by him or her as the ruler's executive branch.

Within each class of Montesquieu's three forms of government there are common general characteristics, but in different societies a wide range of differences in customs and basic principles of governance is superimposed on them. Montesquieu is not satisfied with an arbitrary classification of the diverse forms of society. What makes Montesquieu's approach to the understanding of society so remarkable is his attempt to find the causes, the multiple specific factors, climate, size of nations, traditions, and customs, that first lead to the establishment of the main forms of governance and at the same time introduce variability into each of the three classes of social organization.

By examining the structure of a large number of societies, past to contemporary, Montesquieu attempted to identify as many factors as possible that may determine the specific character of a society. Since the number of factors is very great, he realized that it is unlikely that a complete list could ever be developed, or that the same list will apply to all cultures, or that even within a group of common factors their relative importance will remain the same in different societies. (see Merry 1970; Cohler 1988.)

As a result, Montesquieu suggested several correlations between the three main forms of governance and the physical setting of a given society. As an empirical observation, republican democracies are usually the smallest social bodies, monarchies are of medium size, and despotically governed societies are the largest. This may anticipate the need for increasing controls with increased expanse of societies in general. A barren soil requires a cooperative thrift, compatible with republican democracy.[23] Mountains are obstacles to authority and are regions of freedom. Fertile soil gives rise to wealth and monarchy. Despotism develops in wide plains where great empires can spread over large distances (Durkheim 1960, p. 39). Warm climates induce laziness and generate servitude. Extremely hot climates give rise to polygamy and slavery and the attendant ennui leads to the maintenance of the rules of governance. A law-maker must have insight into the physical setting and nature of that society to adopt the most suitable laws. For example, specific laws are required to achieve the frugality necessary for existence in an arid environment. Laws and societies are both imperfect systems and therefore the laws must be adjustable. In short, according to Durkheim's interpretation of Montesquieu: "A country's type of society, laws and institutions can be deduced from the size of its population, its topography, climate and soil" (p. 40).

To Montesquieu the earth is not just an inert mass following Newtonian laws in its path around the Sun; instead, it is a planet that on its surface supports human existence in a wide range of associations. Although Montesquieu still considers the general aspects of social reality, their relevance is greatly diminished. The specificity of sociopolitical characteristics acquires a predominant importance. It was Montesquieu's intention "to specify the numerous and complex ways in which the political and social systems interact" (Richter 1977, p. 99). "Constitutionalist theories must consider both ideals and the actual working of institutions as affected by all operative forces in the society" (p. 86).

This bare outline may suffice to suggest that Montesquieu is perhaps the first among the great thinkers to seek conceptual continuity in the effectiveness of specific factors as a form of understanding complex sociopolitical reality.

Burke's Interpretation of the English Constitution
as a Complex Sociopolitical System

In England a complex form of constitutional government had come into existence in 1688. Its own revolutionary upheaval, preceding France's by one century, had passed with a transient social disruption. But now the continuity of traditional constitutionalism in England seemed disturbingly threatened by the revolutionary message of the French Revolution from across the channel. At this point Edmund Burke felt summoned to become both an outspoken critic of the French revolution and an unwavering defender of the English constitution (see O'Brien 1992; Ryan 1992).

Burke was born in Dublin, Ireland, twenty-five years, about one generation, after Locke's death; his life spanned from 1729 to 1797. He began his professional career as an author of minor philosophical treatises and as editor of the *Annual Register* for one of the leading publishing houses in London. Burke's growing interest in ethics and politics steered him into direct involvement with public affairs. He entered the House of Commons in 1765 at age thirty-seven and he became an outstanding parliamentarian in the Whig party. In this role he consolidated his perspective on the English and European political scene. His early years in Ireland left in him at least a residual attachment to Irish Catholicism; that, added to his loyalty to English Protestant constitutionalism engendered in him a tinge of sensitivity toward the aspirations of other nations or societies. It became Burke's task to reconcile Locke's and Rousseau's lofty ideas on the rule of the majority with the realities of everyday governance, and to reconcile as well the concepts of liberty and freedom with the occasional rampages of the London mob during the eighteenth century. As Cobban reflects: "To reckon Burke's position as a retrogression would be a mistake; his theories were part of the inevitable deepening and broadening which was bound to follow on the brilliant but superficial guesswork of Locke" (1960, p. 65).

Burke observed the outbreak of the French revolution in 1789 from his vantage point as a constitutional parliamentarian, denouncing and rejecting it one year later in his famous *Reflections on the Revolution in France* (1815, 1969). In a sharp break with the system of thought that had led to the French Revolution, Burke questioned the adequacy of abstract, simplifying political theories for the resolution of problems of governance. Agreeing with Montesquieu's position, he maintained that those problems must be resolved by taking into account, as completely as possible, the specific circumstances that bear on particular political situations. Burke felt that abstract theories lead to extremes of political action—and reaction—and are bound to generate unmanageable disruptions of the social fabric. The following abstract from the *Reflections* indicates the general aspects of Burke's political reasoning (Burke 1969, p. 152). It is included at this point because it refers to the "science of government" and opens an opportunity to see whether Burke's science of social systems has any relationship to the science of ideal and nonideal nonsocial systems.

> The science of constructing a commonwealth, or renovating it, or reforming it is, like every other experimental science, not to be taught *a priori*. Nor is it a short experience that can instruct us in that practical science; because the real effects of moral causes are not always immediate; but that which in the first instance is prejudicial may be excellent in its remoter operation; and its excellence may arise even from the ill effects it produces in the beginning. The reverse also happens; and very plausible

schemes, with very pleasing commencements, have often shameful and lamentable conclusions. In states there are often some obscure and almost latent causes, things which appear at first view of little moment, on which a very great part of its prosperity or adversity may most essentially depend. The science of government being therefore so practical in itself, and intended for such practical purposes, a matter which requires experience, and even more experience than any person can gain in his whole life, however sagacious and observing he may be, it is with infinite caution that any man ought to venture upon pulling down an edifice which has answered in any tolerable degree for ages the common purposes of society, or on building it up again, without having models and patterns of approved utility before his eyes (Burke 1969, p. 152).

In our own time, for example, we can think of the U.S. civil rights movement as one that initially provoked social disturbance but which proved beneficial in the long run. Conversely, the removal of our economic regulatory controls in the 1980s seemed to stimulate the economy initially but it eventually produced the debacle of the collapse of many savings and loan institutions. The unpredicted major changes in the politics of communist Eastern Europe also indicates the existence of the latent unrecognized political forces of the type Burke referred to in the preceding quotation.

Here we can recognize some of the basic tenets of Burke's approach to understanding sociopolitical reality. Burke speaks of a science of governance: but what kind of science was he thinking about? Apparently it was a practical guide to the maintenance of government, based on the gathering and recording of political cause-and-effect relationships that develop into the tradition of a political system, and into part of a constitution. It is a system in which the diverse sociopolitical entities—branches of government, political parties, local administration, individual citizens—share power, identify areas of common interest, agree on compromises, and become in this sense a system of specific forms of conceptual continuity (see Chapter 5).

Burke emphasized that his political systems are more complex, and the cause-and-effect relationships more intricate, than in mechanistic physics. The fact that in his evaluations of the French Revolution Burke drew a sharp distinction between that revolt and those in the United States, Ireland, Corsica, and India indicates the importance he gave to the specific circumstances that came into play in each political situation. Simplified representations of government are defective because they define only parts of the totality of political systems. The greater the number of factors, and the more detailed the description of circumstances that have some bearing on some particular political event, the greater the likelihood for resolution of political conflicts. Actually, Burke holds that at any time it is virtually impossible to recognize all of the factors playing a role in any particular political event. This may be one of the reasons Burke rejected the pertinence of rationalistic thinking and a priori abstractions as a basis of governance. In this sense, political thinking could not be regarded as a science like physics, the only science truly regarded as one in Burke's time.[24]

A specific example of what Burke thinks are the inadequacies of sociopolitical simplifications can be found in one of his comments to a fictitious representative of the French government (Burke 1969, p. 122):

> In your old states you possessed that variety of parts corresponding with the various descriptions of which your community was happily composed; you had all that combination and counteraction, which, in the natural and in the political world, from the reciprocal struggle of the discordant draws out the harmony of the universe. These opposed and conflicting interests, which you considered as so great a blemish in your old and in our present constitutions, interpose a salutary check to all precipitate resolutions: They render deliberation a matter not of choice, but of necessity; they make all changes a subject of *compromise* which naturally begets *moderation;* they produce *temperaments* preventing the sore evil of harsh, crude, unqualified reformations; and rendering all the headlong exertions of arbitrary power, in the few or in the many, for ever impracticable. Through that diversity of members and interests, general liberty had as many securities as there were separate views in the several orders.

Here Burke accepts the conflict of diversity as a given fact of sociopolitical reality; one that cannot be eliminated by disregarding one or another of the opposing tendencies. It becomes a matter of adjusting the relationship of the individual with the state.

Even though Burke made it his main aim to separate himself from the Philosophes by emphasizing the requirements of specificity, diversity, compromise, and moderation, he had to base the structure of governance on certain general elements or principles. In attempting to define a form of government compatible with those requirements, Burke developed an inventory of elements that included "Nature," in particular the conformity and the method of "Nature." Then there are "Principles," "Laws," "Liberty," and "Freedom." In defining these elements, Burke's overall conception of the process of governance became more meaningful.

In contrast to Locke's "Law of Nature," Burke followed the Christian dogmatic interpretation of the concept, much like the thoughts of St. Thomas cited in Chapter 1. Burke recognized a law of God "by which we are knit and connected into the eternal frame of the universe, and out of which we cannot stir" (Cobban 1960, p. 42). The law of God must be regarded as "the great primeval contract of eternal society, linking the lower with the higher natures, connecting the visible and the invisible world" (Cobban 1960, p. 42, quoting Burke). This contract cannot be understood or codified in general rational terms, and its reality is manifest in the persistence of human association, in the traditions that hold societies together, and in the feeling of overall relationship among human beings.

Burke's general law of God (the "Natural Law" of Locke and Rousseau) must be interpreted in specific terms in a second set of laws, directly applicable to the human situation and represented in the specific laws of diverse human societies (see Cobban 1960, p. 43). The law of God is perfect, but its contents cannot be directly and specifically applied to concrete human situations. In contrast, the codified laws of man give concrete and specific rules, but are not perfect. These specific rules can then be justified on the basis of more general social requirements. For example, we can ask whether it is a threat to the persistence of human association. Thus, there is a higher level of social requirements as the basis for persistence and continuity, and a second set of social requirements for the specific implementation of the former. Burke links the two sets of requirements, justifying specific ordinances via the authority of the general demands. It is appropriate to recall once again that Plato and Aristotle also envisaged a body of abstract rules for the maintenance of society in general, and another specific set of ordinances applicable to particular social situations (compare Chapter 1).

Metaphorically, Burke's set of general requirements can be seen as the grammar and his specific ordinances the novels and poems of social legalistic relationships. By anchoring the specific ordinances to a general and theistic body of divine law, Burke precluded the possibility of arbitrary and radical changes in the specific rules, and provided governance with a conservative basis; yet, Locke's system has no superhuman frame of reference, and therefore it can be altered with the abruptness exemplified by the French Revolution. As Cobban has pointed out (p. 43) , Burke started from divine obligation imposed by the law of God, whereas Locke started from "Natural Law" decreed by man himself.

Burke's societal structure included a further set of laws regulating the demands of individuals, a set encoded in their "rights." Burke replaced the abstraction of human rights with a representation of concrete human demands which he saw in the marketplace, or in his dealings with other humans. He tries to reconcile empirical and abstract representations of human rights by suggesting: "Politics ought to be adjusted not to human reasoning but to human nature of which reason is but a part and by no means the greatest part" (Cobban 1960, p. 77). This contrasts with the position taken by most political thinkers of Burke's time, who regarded rationality, in its capacity for abstraction, as the essence of human nature. Cobban adds: "If (the abstract rationalists) are to be admired for upholding an ideal, they are to be condemned for taking the self-constructed ideal for the whole reality" (1960, p. 78).

Again and again Burke emphasized that "rights" cannot be defined as absolutes. In his definition, "rights" are not metaphysical but moral categories. He stresses the precarious balances and compromises between different demands that must be arrived at in almost quantitative terms: how much of the one and how much of the other to provide the most satisfactory distribution of "rights" between contending interests. Modern examples are the contemporary disputes between labor and management when contracts are up for renegotiation.

> The pretended rights of these theorists are all extremes; and in proportion as they are metaphysically true, they are morally and politically false. The rights of men are in a sort of middle, incapable of definition, but not impossible to be discerned. The rights of men in governments are their advantages; and these are often in balances between differences of good; in compromises sometimes between good and evil, and sometimes, between evil and evil. Political reason is a computing principle; adding, subtracting, multiplying, and dividing, morally and not metaphysically or mathematically, true moral denominations (Burke 1969, p. 153).

Burke defined freedom and liberty in terms of the relationship of the individual to the state. The main problems to be resolved, the main dangers to be avoided, are first the totalitarian dominance of the state over the individual, in which the individual activities are nearly fully prescribed by an arbitrary set of rules of human origin. The second danger is that under conditions of unlimited freedom in which social responsibility is lacking, individuals become predator and prey, with the eventual breakdown of any communal system. Both these extremes are simplifications of political reality unacceptable to Burke and incompatible with his requirement for a consideration of the greatest possible number of circumstances and factors in defining a satisfactory individual-state relationship. The answer for Burke lies in those forms of government in which the relationship of individual and state has been maintained over the longest time and with the least possible disturbance of order. Burke indicated the desirable balance of freedom of action and social restraints in a letter to DuPont (Hoffman and Levack 1949, p. 279):

> Liberty is not solitary, unconnected, individual, selfish liberty, as if every man was to regulate the whole of his conduct by his own will. The liberty I mean is social freedom. It is that state of things in which liberty is secured by the equality of restraint. A constitution of things in which the liberty of no one man, and no body of men, and no number of men, can find means to trespass on the liberty of any person, or any description of persons, in the society. This kind of liberty is, indeed, but another name for justice; ascertained by wise laws, and secured by well-constructed institutions. I am sure that liberty, so incorporated, and in a manner identified with justice, must be infinitely dear to everyone who is capable of conceiving what it is. But whenever a separation is made between liberty and justice, neither is, in my opinion, safe. I do not believe that men ever did submit, certain I am that they never ought to have submitted, to the arbitrary pleasure of one man; but, under circum- stances in which the arbitrary pleasure of many persons in the community pressed with an intolerable hardship upon the just and equal rights of their fellows, such a choice might be made, as among evils. The moment will is set above reason and justice, in any community, a great question may arise in sober minds in what part or portion of the community that dangerous dominion of will may be the least mischie- vously placed.

Burke continues this letter by enumerating all the specific liberties that a governing body must safeguard to secure constitutional continuity irrespective of whether that body is a monarchy or a democracy. At the same time, the governing body must abstain from any or all of a long series of infringements on the freedom of its citizens, also enumerated in the same letter.

After stating the basic elements of governance, Burke demonstrated their adequacy by describing their implementation in the constitution of the English government. It is remarkable that Burke did not present us with a status quo of constitutional principles but instead he spoke of reformations and reaffirma- tions of traditional concepts: "All the reformations we have hitherto made, have proceeded upon the principle of reference to antiquity; and I hope, nay I am persuaded, that all those which possibly may be made hereafter, will be carefully formed upon analogical precedent, authority and example" (Burke, 1969, pp. 117–118).

Burke wanted to avoid disruptive discontinuity, by insisting on a persistent frame of reference. Reformation must have a close tie to the past. In this sense Burke pointed out:

> Our oldest reformation is that of the Magna Charta. You will see that Sir Edward Coke, that great oracle of our law, and indeed all the great men who followed him to Blackstone (author of a treatise on the Magna Charta, Oxford, 1759) are industrious to prove the pedigree of liberties. They endeavour to prove, that the ancient charter of the Magna Charta of King John, was connected with another positive charter from Henry I, and that both, the one and the other were nothing but more than a reaffirmation of the still more ancient standing law of the kingdom. In the matter of fact, for the greater past, these authors appear to be in the right, perhaps not always; but if the lawyers mistake in some particulars, it proves my position still the more strongly because it demonstrates the powerful prepossession toward antiquity, with which the minds of all our lawyers and legislators have been always filled; and the *stationary policy* of this kingdom in considering their most sacred rights and franchises as an inheritance. . . .
>
> In the famous law of Charles I, called the "Petition of Right", the parliament says to the king your subjects have inherited this freedom, claiming their franchises not on abstract principles as the rights of man but as their rights of Englishmen, as a patrimony derived from their forefathers.

This position is further reinforced in Burke's letter to DuPont (Hoffman and Levack 1949, p. 294):

> You will observe that, from the Magna Charta to the Declaration of Rights, it has been the uniform policy of our constitution to claim and to assert our liberties as an entailed inheritance derived to us from our forefathers, and to be transmitted to our posterity as an estate specially belonging to the people of this kingdom, without reference whatever to any more general or prior right. By this means our constitution preserves a unity in so great diversity of its parts. We have an inherited crown, an inheritable peerage, and a House of Commons and a people inheriting privileges, franchises, and liberties from a long line of ancestors.

Again Burke rejected abstract principles as the basis of "rights" and emphasized the role of a common inheritance in the unification of a group of people and as a form of conceptual continuity. This is the initial step in the transformation of the diversity of individuals into the unity of a nation, a concept novel in Burke's time.

Regarding tradition as the main force maintaining the relationship between individual and society, Burke invoked an omnipotent providence as its source, that is, an expression of God's will manifest in customs, laws, and institutions (Cobban 1960, p. 85). In Burke's view, traditions, including their continuity and gradual evolutionary change, are aspects of human nature. Here again it is important to restate that all of human nature itself is a product of God's creativity and not an empirically or deductively derived expression of the consensus. Burke's basis for the development of social systems is the obligation of the individual to God's prescriptions whereas to Locke, the concept of God is derived from the natural laws of physics.

Apparently, Burke envisaged a persistence of the whole of society that included a gradual replacement of its parts. Thus, possibility for reformation is opened, and includes continuity of tradition again as a form of conceptual continuity as the primary requirement for the maintenance of a social system. Tradition is an isolated concept, and its relevance to government is not directly explained. In Burke's mind, the existence of a particular tradition means that the relationships between individuals and the whole of society were working satisfactorily over a considerable length of time, certainly for longer than a generation. It means that the regulations restraining individual liberty are not excessive, and are sufficient to guarantee the coherence of the social whole. This type of tradition establishes an empirical raison d'etre for the individual members of a particular society and gives rise to the laws maintaining society. The totality of relationships elaborated in the history of a particular society generates an awareness of traditions that for Burke is the basis and essence of nationality. His concept of nation does not extend beyond itself, however, it is still free of the stigma of assertive sovereignty or of the aggressiveness that became the hallmark of nationalism in the nineteenth century. By introducing the concept of nation as the equivalent of the complexity of traditional relationships, Burke wanted to avoid the polarization that may lead to ideologies. Whether Burke himself was thinking of supra- and inter-national relationships other than in terms of domination and exploitation is left open at this point.

Burke's main work, his *Reflections on the Revolution in France* (1790; Burke 1969), is as much a condemnation of the destructiveness of this social upheaval as it is a commendation of the constructiveness of the English constitution. It is as if Burke were saying that here is a form of social organization that has maintained, with limited aberrations and overall continuity, a working relationship between the citizen as individual and the citizen as a member of the commonwealth. It is also a constitution under which for centuries the individual was not alienated from its "self" and at the same time remained aware of its role as member of the commonwealth. It is an organization that maintains checks and balances of power between an executive arm, the king, his cabinet and his appointed officials; the bicameral legislative arm of the House of the Commons and the House of the Lords; represents through the political parties the interests of the commoners and aristocrats; and includes a corrective judiciary branch that controls the legitimacy of governmental functions. In addition, there are numerous associations of professionals and merchants that can exert political influence. It all has developed small step by small step in a continuity of secular and religious traditions, a continuity which, in Burke's view, was ordained by God. We can readily realize that in defending the English constitution, Burke actually defends the validity of complexity as a primary condition of social reality and conceptual continuity as a form of understanding.

The newly established convergence of terrestrial and celestial mechanics was so successful in unifying and simplifying a large part of physical reality that it suggested its adoption as a rational basis also of sociopolitical reality. The emergence, with the introduction of Galilean-Newtonian mechanics, of a new perspective of the physical universe became a compelling reason to seek a parallel, general interpretation of social organization. The Enlightenment is unique; neither before its time nor hereafter did the thinking about the ideal systems of science, represented here by mechanics, and that about social utopias converge more closely. For a short historical moment, revived only in Marx's philosophy, the hope arose that sociopolitical problems could be solved with the same certainty and predictability as the problems of mechanistic physics. That this expectation proved to be illusory in several respects was the tragedy of the Enlightenment. The reasons why these expectations could not be fulfilled did not immediately become evident. In this sense the Age of Enlightenment did not solve, but did define the problems Western society had to deal with in the subsequent centuries. Some of the inadequacies in the thinking of the Enlightenment are reiterated here as points of departure for discussions in the subsequent chapters.

Although the introduction of gravitational force as a unifying principle of physical reality was a fundamental advance the mode of action of this force, for example, on free falling bodies or the planets, remained unknown to the present time and is still somewhat problematical. In this sense gravitational force and the masses acted upon remained discontinuous concepts. This discontinuity was not abolished by postulating the existence of an "ether" as a mediating agent without evidence for the mode of its hypothetical effectiveness. The full meaning of conceptual continuity became apparent with the establishment of certain ideal model systems of physics that suggested, a least tentatively, conceptual continuity of gravitational attraction as well (Chapter 3).

The thinking about the ideal systems of physics did not even suggest the need to deal with complex forms of reality—inanimate, biological, or sociopolitical—nor did it conceive of any other way of achieving understanding except through the establishment of the broadest possible generalizations as the basis of conceptual continuity.

Actually understanding in terms of conceptual continuity differs in fully defined, ideal systems of physics and in not fully defined, complex systems. In the former category, very general elements establish continuity; in the latter category, which includes complex physical, biological, and sociopolitical systems, highly specific elements are recognized as mediators of continuity. This difference is discussed in Chapters 3 and 4.

The Age of Reason spawned a full range of models for social organization. For some of the eighteenth-century thinkers, mechanistic physics appeared as a well-tended landscape into which human existence should be integrated as part of an overall ordering principle. Thus, Hobbes and the French Philosophes unreservedly adopted a mechanistic approach for the representation of sociopolitical reality. Locke and Condorcet followed only with some reservation this line of thought. The necessity of considering complex, specific conditions in sociopolitics was advocated toward the end of the Enlightenment by Montesquieu and Burke. Rousseau, rejecting mechanistic modeling developed a simplifying utopia, assuming a position between the two groups.

The Age of Reason did not indicate how these models would translate into sociopolitical reality nor did it give a clue whether and in what sense conceptual continuity is established in different societies derived from these models. This is the theme developed in Chapters 5 and 6.

Chapter 3 Conceptual Continuity
in Ideal Physical Systems

The advances of physics in the past two centuries are common knowledge. Nevertheless, some examples of ideal physical systems are included here to illustrate simple forms of reality; to emphasize the establishment of conceptual continuity as an essential part of understanding; to show that many steps are required to arrive at the point where conceptual continuity is recognized; and to demonstrate that different forms of conceptual continuity distinguish ideal and nonideal systems. Very general elements establish continuity in ideal systems, whereas specific elements establish continuity in nonideal systems. This chapter deals with general forms of conceptual continuity; specific forms are considered separately in Chapters 4 and 5.

The simplifying unification of physical reality rapidly advanced with the investigations of thermal processes, light, and other forms of radiation, electricity, and magnetism. It was recognized that mechanical work, heat, radiation, and matter itself were interconvertible forms of energy. The explorations of atomic structure, elaborated by quantum mechanics led to the unification of many physical and chemical phenomena. At the present time, attempts to establish the relationship between quantum mechanics and relativity theory are aiming at an ulti-mate unification of physical reality (Margenau 1950; Blanpied 1969; Orear 1967; Cohen 1985; Harre 1986; Weinberg 1993).

THE CONVERSION OF MECHANICAL WORK INTO HEAT

One of the fundamental advances in the unification of physical reality was the demonstration of the interconvertibility of all forms of energy and the recogni-tion of the principle of the conservation of energy, with the conversion of mechanical work into heat as a prime example (Smith 1990). The fundamental experiment that demonstrated the convertibility and quantitative equivalence of mechanical and thermal energy was carried out by the Englishman James Prescott Joule (1818–1889) in 1843. In Joule's experiment a pulley was turned by a suspended weight. The rotation of the pulley was transmitted to a paddle-wheel in a thermally well-insulated volume of fluid. The energy of the descend-ing weight was directly transmitted to the paddle-wheel and from the wheel to the fluid. In the fluid, the received energy was transformed into the kinetic energy of fluid turbulence, with a concomitant increase in fluid temperature equivalent to the kinetic energy of the descending weight that activated the pulley. In later converse experiments, increasing the temperature of a gas in a cylinder, closed by a moveable piston, was found to increase the gas pressure in the cylinder and to move the piston, demonstrating the conversion of heat into mechanical work.

Joule's experiment acquired great prominence because it was followed by the realization of the interconvertibility of all energy forms. This led to the recogni-tion of the fundamental principle of the conservation of energy, promulgated in 1847 by H. L. F. von Helmholtz (1821–1894). The principle states that the total sum of energy in the universe is constant. Energy cannot be created or de-stroyed, and energy can only be converted from one form into another. As a further extension of this principle, it became evident that during all types of energy conversions some heat is lost in a form that cannot be used to generate any work. It was thought that eventually all forms of energy in the universe are dissipated as useless heat, and dissociation of organized matter into an inert, homogeneous state (Smith 1990). These advances not only led to one form of unification of physical reality of the nineteenth century but also established the basis for the industrial development during this period.

Similar to the Newtonian unification here again pivotal advances were made without a corresponding advance in clarifying the conceptual continuity between the different energy forms. To come back to Joule's experiment, the conversion of mechanical energy into heat, although experimentally demonstrated, quantitatively defined, and mathematically formalized, the actual steps in the transformation of mechanical work into heat remained obscure. Initially heat was thought to consist of a separate component, called the "caloric," inherent in all matter. This "caloric" would dissociate and appear as heat when the state of matter was mechanically disturbed. The strict dependence of heat production on the amount of work done, irrespective of the material and the form of work involved made such an interpretation untenable. During Joule's time, physicists began to understand that the molecular composition of matter suggested a hypothetical role of molecular motion in the transformation of mechanical work into heat as the basis of conceptual continuity between the two energies. The conclusive evidence for molecular motion as an energy mediator was established through the later developments of the kinetic theory of gases and the advances of statistical mechanics of molecular energies. Through these developments, molecular motion as mediator of energy conversion became one of the most general forms of conceptual continuity.

LIGHT AND THE QUANTUM MECHANICAL ESTABLISHMENT OF CONCEPTUAL CONTINUITY OF RADIATION AND MATTER

The interaction of light and matter is considered here in some detail as one further step in developing the idea of conceptual continuity in physical systems. The interaction of light and matter is suggested by many forms of everyday experience: the tanning effect of sunlight; the chemical effects of light, for example in photography; the requirement of light for the normal growth of plants; the conversion of solar radiation into other forms of energy, such as electricity or heat. We take for granted the reflection of light in a mirror, its refraction by our eyeglasses, or its diffraction through raindrops into a rainbow. These experiences tell us that light must interact with matter in some way to produce these phenomena.

Again, qualitative observations and quantitative definitions of these phenomena are the first two steps in the exploration of the nature of light. They constitute optics, a field that developed into its modern state during the past four centuries. The three main types of interactions between light and matter first investigated were reflection, refraction, and diffraction. One of the first basic correlations made was that in reflection of light from a surface, the angles between the incident and reflected light and a perpendicular to the surface are the same. In refraction a light beam is bent as it passes from one medium to another, for example, from air into water or glass. The ratio of the angles between the light beams in the two media and a perpendicular to the boundary between the media is the refractive index, which remains constant for all angles of the incident beam.

Diffraction consists of the bending or spreading of light as it passes through a narrow opening in an opaque diaphragm. The spreading of diffracted light, passed through two narrow slits placed close together, overlaps to form alternating light and dark bands on a screen, a phenomenon known as interference. Interference was first explained by assuming that light was propagated through the slits in the form of waves. Depending on the angles at which the light left the adjacent slits, the hypothetical waves coming from the two sources were propagated in-phase or out-of-phase. That meant the light waves propagated in-phase had matching crests and troughs, whereas those waves propagated out-of-phase had crests aligned with troughs and(or) troughs aligned with crests. In the former case, the light waves were thought to reinforce each other to form the light bands, and in the latter case, the light waves canceled each other out to produce the dark bands. Thus, interference seemed to support the assumption of the wave nature of light in accordance with Maxwell's interpretation of electromagnetic phenomena. A hypothetical linear propagation of light in the form of particles was not expected to produce the light and dark bands of interference patterns.

The laws of optics given above are among the best examples of quantitative correlations. With their high level of mathematically defined precision, these laws led to the development of important and widely used optical instruments such as microscopes and telescopes. Observations made with these instruments immensely broadened our understanding of life and of the reaches of the universe and continue to do so. The question of how light actually interacts with matter in producing these optical phenomena remained unanswered at first. Light and matter remained at this point conceptually discontinuous. Establishment of conceptual continuity between light as well as other forms of radiation and matter depended on the dramatic developments of quantum mechanics in the first part of the twentieth century. These developments demonstrated that all forms of electromagnetic radiation, including light, consist of elementary energy units, or quanta, designated as photons. Quantum mechanics established the basic principles of the interactions between photons and electrons, one of the elementary units of matter. This development is now considered in more detail to show the long series of steps that are sometimes needed to represent initial mathematical relationships in the form of conceptual continuity.

Nineteenth century investigations led to the conclusion that each of the elementary forms of matter was composed of a specific kind of atom, and that some groups of elements shared certain physical and chemical properties. The assignment of the various elements to those groups gave rise to what the Russian chemist, Dimitri Mendeleev (1834–1907) called the periodic table of elements in 1869. Eventually more than a hundred elements were identified corresponding to specific atomic structures. For a time the multiplicity, apparently inherent in the atomic concept of matter, seemed to be quite far from the ideal simplicity of the physical representation of reality. Only the introduction of the atom theory by Nils Bohr (1885–1962), and its subsequent elaboration in quantum mechanics began to substitute a general, simplified concept of atomic structure.

Of crucial importance to the development of the atom theory was the observation that heating diluted hydrogen gas and examining the light it gave off, produced three sets of distinct spectral lines. The difference in the light energies corresponding to these lines could be expressed as basic energy units, and confirmed the idea that these energy units are the smallest amounts of energy that can be transmitted or absorbed. According to this interpretation, radiation energy cannot be reduced indefinitely but exists in the form of "packages" of very small but definite quantities—the quanta. The recognition of energy quanta and of the quantal differences represented by the energy levels of the hydrogen spectral lines became the basis of Bohr's atom model.

Bohr postulated that in the hydrogen atom a negatively charged single electron circles the much heavier atomic nucleus just as the earth circles the sun. In its ground stage (baseline or normal state), the electron does not emit light or any other form of energy. When hydrogen gas is heated, energy is absorbed by its electrons, which are then lifted to higher-energy paths around the nuclei. As the electrons return to their ground levels, the same energy is released in a new form: photons of ultraviolet or visible light. Since the amount of energy, absorbed or emitted, cannot be smaller than one energy quantum, the energy of the wavelength of the emitted light also must differ by the discrete energy values required to produce the separate spectral lines. In Bohr's model of the hydrogen atom, the negatively charged electron circles the positively charged nucleus only in discrete orbits corresponding to specific energy levels. In atoms of other elements, only those electrons that differ at least in one orbit-determining parameter can share the same orbit. Electrons may differ in the direction of their spin around their own axes or in the shape or orientation of their respective orbits. This is known as the Pauli exclusion principle, named for its originator Wolfgang Pauli (1900–1958).

The later mathematical definition of the state of electrons in the atom by Erwin Schrödinger (1887–1961) led to their representation as standing waves. This appeared to account for the discreteness of the electron energy levels within the atom, since only electron orbits with integral (whole) wave numbers were possible. It also suggested that elementary particles—electrons, photons—could be maintained in a state of undulatory wave motion, and that radiation itself may have the properties of both waves and particles. The question of how electrons are kept in motion around the nucleus remained unanswered at that time.

Bohr's atom model formed the basis of the next advances in the understanding of the interaction of light and matter. One of the decisive steps was the prediction by Albert Einstein (1879–1955) of the photoelectric effect and its subsequent experimental demonstration by Robert Millikan (1868–1953). Millikan demonstrated that a current of electrons was produced by light of sufficient energy (essentially in the short-wavelength, ultraviolet range), when the light struck the surface of a metal plate serving as the negative pole of an electric circuit (Blanpied 1969, p. 351). Energized negatively charged electrons were ejected from the metal, collected on the positive pole of the circuit, and the level of the electron flow was measured. With increasing light intensity, increasing numbers of electrons were released from their static place in the metal plate, and the measured electric current increased proportionately. Release of electrons and the corresponding current were suppressed by a negative potential inserted into the circuit. The inserted negative potential required to obliterate the electron flow was the same, irrespective of the light intensity. This showed that the energy transferred to the electrons by the photons of the incoming light had a finite value that corresponded to the energy quanta of photons.

The release of electrons by absorption of light energy in the form of single photons can be followed in a photomultiplier. This instrument converts a light-induced electron flow, even the incidence of single electrons, into an audible signal. At an ordinary incident light intensity, electron flow in the photomultiplier tube is transformed by the apparatus into a continuous sound. Lower light intensities produce a rapid succession of clicks, and at minimal light intensity the clicks become separated by long time intervals, each click indicating the incidence of a single photon on the collecting element of the photomultiplier. When the photon strikes a single atom on the collector, it activates one of its electrons, inducing a current in the apparatus, which results in the single click. This demonstrates the discontinuous, quantum nature of the incident light. In contrast, if light waves were continuous, they would generate a continuous sound of lower intensity. The occurrence of separate clicks corroborated the idea of light as consisting of photons, carriers of discontinuous energy quanta. At the same time these observations clearly demonstrate the interactions between the elementary units of light and matter and establish conceptual continuity between these two previously discontinuous physical entities.

In a converse experiment, electrons are ejected from a hot, electrically heated metal filament and are further accelerated by applying an electric potential of several thousand volts to their paths. The electrons impinge on the positive pole of an electric circuit, and part of the electrons' energy is absorbed by the atoms of the receiving metal, increasing their thermal agitation. The metal heats up and glows; the glow consists of light from the entire visible spectrum, and also includes x-rays. This effect demonstrates that radiation in the form of electrons has been converted into radiation in the form of photons with measurable energies that increase from those of visible light to those of x-rays. But whether photons themselves are massless energy quanta or are truly some form of matter has not been answered with certainty. Under certain conditions, however, collisions of photons and electrons produce an effect similar to the scattering of colliding particles, indicating some association between photons and the state of matter.

At this point we are left with the often stated ambiguity of light: it is either propagated in the form of waves in a field continuum, as conceived by Maxwell as defined by the laws of classical optics (reflection, refraction, diffraction), or as a stream of discontinuous energy quanta in the form of photons. An elaboration of quantum mechanics known as quantum electrodynamics (QED) led to a resolution of this apparent dichotomy and established further support for conceptual continuity in the interaction of light and matter as suggested in the following outline (Morris 1987; Feynman 1988).

The phenomenon of reflection is used here in indicating the basic differences in the interpretation of light in the classical theory of electromagnetic radiation versus its interpretation under quantum electrodynamics. Understood as electromagnetic radiation, light travels from its source as a bundle of waves, called a wave bundle, to a reflecting surface. Without giving a precise mechanism here, it has been proposed that at least some of the incident waves continue on the reflected beam. The main argument supporting the wave nature of the reflected light is based on the observation that incident light, reflected by two closely spaced surfaces, for example, the upper and lower surfaces of a very thin glass sheet, show interference. This was described earlier here, for diffraction, which came to be regarded as incontrovertible proof for the wave nature of light.

With the strong evidence for light consisting of photons, quantum electro-dynamics attempts to show that the phenomena of classical optics, such as reflection, can be better defined by suggesting a mechanism for the interaction of photons with the electrons of the reflecting surface. QED postulates that the photons emitted from a light source reach all parts of a reflecting surface randomly. There, the incident photons (quanta) are absorbed and later ejected by the electrons of the reflecting surface material. Only those photons that follow the shortest paths from the source to the reflecting surface and from there to the receptor have sufficiently high probabilities of synergistic timing and path direction to form the reflected light beam. Those photons striking the reflecting surface so as to give longer paths from the source to the receptor will interfere with each other's motion and result in very low probabilities of actu-ally reaching the receptor. We are no longer dealing with a deterministic mechanism of reflection but with probabilities as the basis of the distributions and translocations of photons. Because it may be possible to define optical phenomena by studying random interactions between photons, and between photons and matter, and because of other, extensive independent evidence for the existence of photons as the main constituents of all forms of electromag-netic radiation, the quantum theory of electromagnetism and light is strongly favored over the wave theory.

To conclude this basic discussion of quantum theory, radiation energy quanta in the form of photons are absorbed by elevating the energy levels of electrons in atoms, and the quanta are reconverted into photons of emitted light by the return of electrons to their energetic ground states. The decisive point here is that we achieve conceptual continuity in the sense that both radiation and its effects on the atomic constituents of matter are represented in the same terms of absorption and release of energy quanta in the form of photons. Moreover, exchange of energy, usually regarded as function, and structural change in the atom become interdependent with the unification of structure and function on the atomic level.

Photons can assume widely differing (and possibly high) levels of energy during very short times. The existence of these transient photons states has been inferred from slight shifts in the wavelengths of light that those photons may emit (Morris 1987, p. 56). This class of photons has been termed "virtual photons." Even the generally accepted standard speed of light of 186,000 miles or 300,000 km/second may be only an average figure. During short time intervals, individual photons can travel at speeds larger or smaller than this average figure. The question has been raised recently whether electrons exist as independent particles or as negative charges together with a field of virtual photons.

The preceding discussion supports the idea that the photon itself is the universal energy carrier in electromagnetic interactions at a distance and a link providing conceptual continuity between the interacting elements. Photon exchange is held to be responsible for all electromagnetic interactions, and photon exchange between electrons and the atomic nucleus is even considered as the mechanism whereby electrons are maintained in their orbits. The photon can be regarded as the most general link establishing conceptual continuity between the basic physical entities, radiation, energy, and matter.

THE RELATIVISTIC REVISION OF NEWTONIAN MECHANICS

The preceding section indicates one of the two major systems of thought and experimentation that led to the unification of a large part of physical reality. A further major unification is the revision of Newtonian mechanics by Einstein's Theory of Relativity (Blanpied 1969; Morris 1987). Newtonian mechanics is based on observations of motions of objects in space. As one of the basic tenets of this system, the distances traveled by moving objects and the time elapsing during their movement is the same for all observers. Common sense seems to confirm this: how else would it be possible to obtain agreement in the readings of simultaneous radar trap speed readings taken in front of and back of a speeding car, or to follow and chart the trajectories of satellites from different locations? For velocities many magnitudes smaller than the velocity of light these types of observations seem to give adequate agreement. Our notions of speed and time are assumptions that should hold under all conditions if Newtonian mechanics has universal validity. A reexamination of these assumptions leads to a different conclusion. For example, an observer can measure the time required for a light signal to reach a mirror and return to its place of origin. Time is thus calculated as the distance from the observer to the mirror (distance y) and back again (distance 2y) divided by the speed of light (C_i 300,000 km/second), giving the elapsing time a value of 2y/300,000 km. If the mirror is moving away from the observer, the same time interval should be observed for the round-trip of the light signal, according to Newtonian determinism. But since the light has to travel a longer distance, a constant time interval requires an increase in light velocity in order to cover the greater distance in the same time.

In reality the velocity of light remains constant when emitted from objects either at rest or in motion. But as required by the Special Theory of Relativity, it follows that the time interval required for the light to travel out to the mirror and back again while observing the system at rest $[t_A(a)]$ or in relative motion $[t_A(b)]$ must differ. One can calculate the value of the time interval required for the light to reach an object moving at a velocity of v km/sec and its return to the observer by the equation

$$\Delta t\ (Ab) = \frac{\Delta t\ (Aa)}{\sqrt{1 - v^2/c^2}}$$

where c^2 is usually several orders of magnitude greater than v^2 and the correction is negligible under ordinary circumstances. If the motion of the observed object approaches the speed of light the correction becomes significant, and leads to an expansion of time.

The expansion of time postulated by the Special Theory of Relativity has been corroborated by observations of subatomic particles. Experiments using high-energy accelerators have demonstrated that collisions between high-speed particles of the atomic nucleus (protons and neutrons) produce subatomic particles designated as p and k mesons. These latter particles decay rapidly, to produce particles called muons which decay in an average of 2.2×10^{-6} seconds. Muons were first detected as components of cosmic rays, which are produced by the interaction of protons from space and are bombarding gas molecules of the upper atmosphere at about 30,000 meters altitude. Considering the muons' extremely short life-spans, they should not have enough time to reach the earth's surface and be detected when traveling close to the speed of light, since they will cover the distance from the upper atmosphere to the recording instruments on the earth's surface in only 10^{-4} seconds. The observation of these particles becomes possible by earth-bound instruments (Blanpied 1969, pp. 384–385) only by an expansion of time due to their high speed.

Corresponding to the expansion of time observed when objects move at very high speeds, approaching the speed of light, an increase in mass and a decrease in the length (in the direction of motion) are also consequences of the relativistic modification of Newtonian mechanics. The measurements for objects moving at different relative speeds can be related to each other by a set of mathematical transformations derived by the Dutch mathematician, A. H. Lorentz (1853–1928) in 1903.

A further important aspect of special relativity is the relationship of mass and energy. If two bodies with rest masses of 2Mo collide, an increase in total rest mass is taking place.

$$\text{Total Mass} = \frac{2\text{Mo}}{\sqrt{1 - v^2/c^2}}$$

The increase in mass is caused by the conversion of collision energy into mass, which becomes significant when the masses approach the velocity of light. A generalization of this relationship gave rise to Einstein's famous equation, $E = Mc^2$, the basis for the conversion of matter into energy in atomic reactions (Blanpied 1969, p. 411).

The transformation of the deterministic Newtonian concepts of time, mass, and linear dimensions into flexible relativistic concepts and the convertibility of mass and energy are the main consequences of Einstein's Special Theory of Relativity. In its further elaboration, the General Theory, Einstein postulated that a space-time continuum exists in a curved shape, imparted to it by the masses it surrounds. In this interpretation, gravitational attraction is not a force in the Newtonian sense, but a manifestation of space-time geometry. As in the case of Einstein's Special Relativity, the theoretical abstractions of General Relativity were also corroborated by giving a correct prediction for the orbit of the planet Mercury. The theory also predicted the bending of light in the vicinity of large masses such as the sun, and the slowing of the periodicity of atomic radiation. Furthermore, according to general relativity, space-time should be greatly distorted in the vicinity of masses with extremely high density as in the case of collapsing stars. The changes in the mass distribution during such a collapse should produce corresponding modifications in the conformation of space-time. These space-time changes should generate gravitational waves of an intensity that could be recorded on earth. Continued preparations for the recording of these gravitational waves are in progress, although attempts to obtain evidence for such waves have failed so far. The actual demonstration of gravitational waves resulting from astrophysical upheavals such as the collapse of stars would provide a definitive confirmation of Einstein's general relativity theory (Abrahams and Shapiro 1990). The predictions of particular, observable events has indicated the predictable, determinist nature of the relativity theories. Thus, Einstein's unique genius created a form of physical reality in which space-time and gravitational attraction led to a conceptual unification which also includes interaction at a distance without any requirement for an ether as carrier of mass gravitational interactions.

This chapter gives a brief outline of three major advances in the unifying simplification of physical reality that led to very general forms of conceptual continuity. What characterizes physics and makes it so remarkable is the completeness with which it replaces, by general abstractions, the diversity of phenomena and the specificity of the objects it describes. These abstractions are the "ideal systems" of physics. The extent and rapid pace in the advances of the understanding of the ideal systems of physics and their practical applications gave rise to the unqualified acceptance of the superiority of simplifying generalizations and their application to the resolution of the problems of nonideal systems including human society.

The ideal systems, their mathematical formalization, and the corresponding simplifying unification of the universe exist only in our minds. Nowhere in the universe can we find these ideal systems. But, the "nonideal" systems we encounter in our daily routine, whether they be technological, biological, or sociological, are manifestations of different degrees of specificity and complexity. That does not mean that these nonideal systems are not compatible with the generalizations of physics. In fact, they all are. (See the opening statement of the Introduction.) The generalizations are merely not sufficient for the understanding of nonideal systems. The main point here is a reevaluation of the relevance of abstraction and generalization used in defining the ideal system of physics for the understanding of nonideal inanimate systems of geology or meteorology, the complexities of biological systems, or the sociopolitical changes in the overall course of history. This question is examined later in the book.

Chapter 4 Complexity as the Reality of the Postmodern World: Conceptual Continuity in Nonideal Systems

The simplifying unification of physical reality is a characteristic of the modern phase (about 1700–1950) of Western civilization. The recognition of complexity as an attribute of reality is here considered as a characteristic of the postmodern phase (Toulmin 1990), the period following the Second World War. The intrusion of complexity into the perception of reality in postmodernism had three sources. First, the advance of abstract science created a staggering fall-out of practical applications in industry and related developments. Within slightly more than a century the traditional Western society, based on manual farming and craftsmanship, was transformed into a machine-operating social organization. The resulting increase in the supply of agricultural products and manufactured commodities, the improvement in housing and communication, the elimination of diseases, and the spreading of education and literacy in general were regarded as progress. By enhancing possibilities for the expression of the individual within the emerging democratic societies, this progress engendered the increasing complexity of everyday life.

A second source of the perception of complexity was the presentation by the media, in particular television, of the full gamut of diversity of local and world events and of the staggering problems of environmental maintenance, population dynamics, and social adjustments.

The third source of postmodern complexity was a reorientation in scientific thinking. In traditional thought, complexity was a category to be avoided or to be replaced by simplicity. Certainly complexity was not to be investigated in its own right and in all its dimensions. It was a concept that did not find a place even in the major *Encyclopedia of Philosophy* (Edwards 1972), and remained neglected as long as the ideal systems of physics were the only models of reality. This changed in the postmodern era with an increasing concern about nonideal systems. The beginning of this trend was heralded by Warren Weaver's well-known article "Science and Complexity" (1948), which called attention to "problems which involve dealing simultaneously with a sizable number of factors which are interrelated into an organic whole." By referring to wholes comprising a large number of related parts, Weaver suggests a meaning for the concept of systems that became the basis of what we now call systems theory (Bertalanffy 1960; Klir 1972). The hierarchical organization of systems was later on emphasized by Simon (1962). Taking information as a system of related parts Shannon and Weaver created information theory (1949; see also Papentin 1980). Later studies began to define "unstable aperiodic behavior in nonlinear dynamic systems" (Kellert 1993), subsumed under the heading of chaos theory (Gleick 1987). The development of powerful computers provided the methodological means for the study of complex phenomena (Casti 1979, 1994; Gottinger 1983; Calode 1988; Flood and Carson 1988; Kauffman 1993). The full scope of complexity as a representation of reality became apparent in a symposium *The Science and Praxis of Complexity* held in 1984 under the auspices of the United Nations University (Aida et al. 1985).[1]

As a basis for further discussion of the nature of complexity in physical, biological, and sociopolitical reality, a survey of some representative systems with the apparent characteristics of complexity follows. This section will focus on the nature of conceptual continuity in nonideal systems with its distinctive significance for the understanding of living matter.

THE LUMPY UNIVERSE

In Plato's cosmology, the universe had the geometric regularity and harmonious symmetry of crystalline spheres. The geometric form was conceptually satisfying and gave the universe the characteristics of an ideal system. Plato's universe became incompatible with Copernicus's heliocentric planetary system, but the idea of geometric perfection of the celestial order was still an important consideration in Kepler's early studies of planetary movement. Only the further advance of astronomy, the increasing realization of the vastness of interstellar distances, and the apparent absence of obvious order in the cosmic distribution of matter led to the abandonment of the Platonic conception of the universe.

The development of a contemporary, comprehensive hypothesis about the genesis of the universe and the distribution of matter in it began with Edwin Hubble's 1929 observation of the rapid, centrifugal movement of galaxies. This suggested that it all originated at one central point some 10–20 billion years ago, in an explosive event in which energy was converted successively into subatomic, atomic, molecular, and supramolecular forms of matter. Planets, stars, and galaxies are still driven apart by the momentum born in the initial explosion. This "Big Bang" hypothesis of the formation of the universe has found much empirical support (see Weinberg 1977; Morris 1987). The hypothesis predicts that part of the energy produced in a Big Bang type transformation should persist as a weak, uniformly distributed radiation. This type of radiation was actually discovered with one of the huge antennae used for satellite communication. The distribution of some elements—hydrogen, helium, and lithium in the galaxies—expected according to the Big Bang hypothesis, was also substantiated. However, the Big Bang hypothesis has not been the only mechanism postulated to describe the genesis of the universe. Alternatively, the formation of galaxies from space with high densities of charged particles (the so-called plasma condition) has been considered as a possible alternative (Peratt 1990); the Big Bang hypothesis is adopted here as a basis for the following comments.

The apparent uniformity in the background radiation in different regions of the universe suggests that only minor local perturbations existed during the initial period of the unfolding universe and that particles, probably in the form of photons, electrons and neutrinos were uniformly distributed at high densities at that time. With the expansion of the universe the particle density was sufficiently reduced to allow small density fluctuations, in particular of neutrinos, which initiated the formation of the galaxies, each with its own characteristics, and the assembly of galaxies into structural aggregations that came to delimit the boundaries of immense extents of space (Silk, Szalay, and Zeldovich 1983). Some galaxies attract others, inducing their movement independent of the expansion of the universe. Currently, evidence has been obtained that such "attractors" also enclose vast quantities of invisible (dark) matter that presumably consists of high-mass, slowly moving particles, currently of poorly defined nature. Other galaxy clusters do not exert such gravitational attraction and seem to enclose empty space without detectable quantities of dark matter (Dressler 1989). Whether the apparent "empty" space is an absolute void or is permeated by energy fluctuations has remained controversial. Also, the possibility of local expansions of the universe, the so-called wormhole hypothesis, has been considered to account for anomalies in such energy fluctuations (Freedman 1990).

Within some galaxies, the collapse of stars that have used up their thermo-nuclear fuel creates regions of extraordinary density. In their vicinity, space-time is highly distorted and induces gravitational fields of such an intensity that they prevent the escape of light or matter. These are the "black holes." Colliding black holes become sources of enormous quantities of radiant energy known as "quasars" (quasi-stellar radio sources). Another type of high density stars, "pulsars," spinning around their own axes, produce radio waves that have been recorded at periodic intervals (wave pulses). Binary pulsars, which are paired quasars, revolving about each other are expected to generate changes in the gravitational field (Abrahams and Shapiro 1990). Taken together, the recognition of the diversity of cosmic phenomena and the irregularity and the distribution of matter has led to the semi-popular term "lumpy universe" (Cook and Schrof 1990).

In recent approaches to understanding the distribution of matter and energy in the universe, attempts have been made to account for local space-time distortions by supercomputer modeling, as an alternative to the usual representation of the universe as a uniform continuum. In changing the image of the universe from simplicity to complexity, the following quotation may be relevant:

> With a supercomputer, the complex set of equations describing the flows of matter and energy and how they interact with space-time can be solved at many distinct points. Even though space-time curvature is continuous, these discrete solutions can provide a useful picture of it, just as the dots on a television screen approximate the continuous shading of reality. The finer the grid points, the more closely the solution approaches reality (Abrahams and Shapiro 1990, p. 33).

Irrespective of how many models it may take to come to a more definitive picture of the genesis and structure of the universe, it is quite evident that the universe, even in the form we perceive it today, has little left of its status as a preeminent example of an ideal system. The arrangement of matter in the

universe bears no resemblance to geometric perfection, and the specific locations of the galaxies are most likely the results of randomly distributed attraction centers and not of a general principle determining the distribution of
matter in the universe. It is still not certain whether and how long the universe
will continue to expand and whether this expansion will be followed by contraction, compaction, and another Big Bang phase of an expansion-contraction
cycle. Also, we cannot predict just when certain stars in specific galaxies will
become extinct and whether their destination will be reversed by some atomic
revival.

A universe without the hallowed pattern of full predictability is perhaps
more awe-inspiring than a universe that carries the human imprint of geometric regularity. At the present time the universe appears to be a loosely organized,
hierarchical structure comprising the levels of individual stars and planets, of
planetary systems, of galaxies, and clusters of galaxies. These units are linked by
unspecific gravitational interactions with the hypothetical interaction of gravitational radiation and matter as a form of conceptual continuity. In such a
representation, the universe is an example of complex reality in which conceptual continuity is still established by the abstract generalizations of mechanics,
quantum mechanics, and the Big Bang theory.

THE VAGARIES OF GEOGENESIS

At the end of Chapter 2 the earth was regarded, purely and simply, as a center of
gravitational attraction, a component of one of the fundamental ideal planetary
systems that have been defined by both Newtonian and relativistic mechanics.
As an alternative, the earth can be perceived as a changing assembly of highly
diverse forms of matter and substances and of coexisting geological, meteorological, and biological phenomena. We have to reorient our determinist,
cause-and-effect thinking from terms of simple collisions of two billiard balls,
moving along straight lines on a low-friction billiard table, to the complexities
of the weather, and of the drifting of continents.

On a world map, the continents are spread out in bizarre shapes, amidst the
vastness of the oceans. Nowhere is there the elegance of symmetry, only the
possibility of a coarse fitting of now-distant coast lines. The specific features of
the different regions were explored, mapped, and recorded in more and more
detail, beginning with the great expeditions of the sixteenth century, and more
and more specific information accumulating over the next centuries. The
charting of the polar regions, the interior of Africa, and other tropical regions

concluded this phase only recently. These explorations of the earth led to tentative conclusions on the origins of geological formations and to an understanding of the geologic eras, their ages, and the major events in each.

The obvious features of the terrestrial landmasses first seemed intractable to further simplification. The geographic patterns were too irregular to be subsumed under a single geometric explanation and not suited for even a statistical analysis. Attempts to answer the question of "How did all this come about?" became an initial approach toward understanding this complexity. As examples, we will follow the genesis of the southern part of South America (Ramos 1989) and of the Tibetan plateau (Monastersky 1990a,b; see also Nance, Worsley, and Moody. 1988).

The starting point of the South American Geogenesis is a mountain range, the Sierra Pampanea in central Argentina. From the types of rocks found in the exposed parts of these mountains, geologists have concluded that about 550 million years ago this region was one of several small separate landmasses, or terrains. These landmass blocks coalesced by repeated collisions. Preceding the coalescence, the ocean floor to the east of the Pampanean terrains had been coursing downward underneath it, a process known as subduction. This sinking added pressure to the hot molten layer of the earth beneath, and gave rise to volcanic activity along the margin of the subduction zone. Also, some surface parts of the sinking ocean floor were scraped off the sinking layer and became part of the Pampanean terrain. A major landmass, the Río de la Plata craton, moving in from the east, produced an uplifting of the Pampanean fragment to form a mountain range perhaps as high as the Himalayas. In the course of time this range was eroded to its present height of 3500 meters.

These local changes were superseded between 530 and 570 million years ago by a general confluence of the continental plates of the earth with the temporary formation of a single supercontinent, Pangea I. A destabilization of this block led then to its breakup into several large continents. One of these was *Gondwanaland*, which comprised the present landmasses of South America, Africa, India, Antarctica, and Australia. Following this period the western edge of the Pampanean terrain became the eastern boundary of an oceanic trough that was sinking below the continent. There again followed volcanic activity in the Pampanean region, along with an accretion of ocean-floor crust that had been scraped off in the course of the subduction process; additional ocean crust was left behind as part of the Pampanean geology. These events occurred between 440 and 550 million years ago. Between 360 and 440 million years ago the accretion of several lengthy, narrow slivers of land occurred on the western edge of Pampanea. Some of these strips moved in from the south, corresponding to present-day Chile; others were translocated from the west with the subduction zone of the ocean floor persisting on the western edge of the newly aggregated continental plate.

In the following 100 million years Gondwanaland was again joined by the other continental landmasses to form the super continent, *Pangea II,* suggesting a cyclic coalescence and dispersion of continental landmasses with the breakup caused by overheating (Nance et al. 1988). This led to the formation (about 240 million years ago) of the continental landmasses roughly in their present shapes, with the eastern coast of South America and the western coast of Africa conjoined. A rift between what was soon to become South America and Africa initiated the separation of these two new continental plates. Recognition of the similarities of the rock formations in the southern, facing parts of the two continents became the basis for the continental drift theory of Wegener (1880–1930), the first to conceive of continental movement as the basis for present geography, and therefore to suggest one of the generalizing schemes of modern geology. The slipping away of Africa from South America's eastern side again initiated subduction in the ocean floor under the western side of the South American continent. The subduction accelerated about 115–135 million years ago, and led to the compression of the western edge of the continent with a horizontal shortening of as much as 240 kilometers at the edge. This compression generated the uplift of the Andes and their volcanic activity.

During the breakup of Gondwanaland, the Indian subcontinent was set adrift and followed a course toward the north, colliding eventually with the Asian landmass. Asia itself was the result of successive accretions of small landmasses, starting from a block in Eastern Siberia about a billion years ago. The area of the Takla Makan desert was added 300–400 million years ago, and at about the same time, fusion with the European landmass took place at what was to become the Ural mountains. At the southern edge of that landmass the ocean floor was subducted, causing folding and volcanic activity that preceded a collision with the Indian shield as recently as 69 million years ago. It was this subduction process that actually carried India northward. The thrust of India's collision with the Asian continent caused the uplifting of the Himalayas, of the Tibetan plateau, and of mountain ranges as far as 1000 kilometers north of the actual fusion zone (Molnar 1989).

The uplift of the Tibetan plateau has been attributed to a direct compression by the impinging Indian landmass and, possibly, to an upwelling of hot, deep material underneath the Tibetan plate with a resulting buoyant uplift of about

1000 meters. An extrusion of Tibetan crust material appears to have occurred first in an easterly and later in a southerly direction. This extruded material formed the geological structures at the boundary between southern China and the Indo-Chinese peninsula (Monastersky 1990a,b). The extrusion pattern is corroborated by laboratory experiments using sheets of plasticine as modeling clay landmasses.

The geogenesis of other regions is similar in principle. The Eurasian landmass represents the accretion of at least ten separate geological units. North America is apparently an aggregate of seven landblocks that became assembled perhaps as long as 2 billion years ago. In the formation of the North American continent, the north slope of Alaska became attached after an extensive translocation from the Canadian territory. At that time, its climate was much warmer than it is today, and its then-rich vegetation became the source of important modern oil fields.

The examples above show that subduction, translocation, collision, and erosion are the main processes modeling the complex features of the earth. The mechanisms of these changes involve the general principles of thermal convection of the deep material carrying the continental plates, and of the mechanical translocation of colliding landmasses as a general and still relevant basis of conceptual continuity. Here, however, the main aim is the identification of the specific circumstances that are responsible for the described terrestrial events. The greater the number of factors that can be recognized to contribute to a final state, the more completely the system can be defined, and the closer understanding approaches conceptual continuity. It seems evident that geogenetic processes are specific instances of classical mechanics and thermodynamics: movement, subduction, and collision of sub-ocean crusts and continental plates can be regarded as billiard-ball-type mechanics on a gigantic scale. The new dimensions that enter into the consideration of geogenesis are the apparent historical uniqueness of the origin of the earth in the random Big Bang dispersion of matter and energy, the earth's location in its galaxy and its position relative to the sun, and the proportions of oceans and landmasses on the earth itself. It is this geohistorical uniqueness that differentiates these complex inanimate systems of nature from man-made machines.

Uniqueness and specificity are presented here as the distinctive aspects of complex, nonideal, inanimate systems. This does not exclude the establishment of some generalizations (such as the theory of continental drift) that aid our understanding of geogenesis. Nevertheless, it is the geogenesis of specific landmasses, such as South America or Tibet, that are the predominant objects of geological investigation. In these studies, the basic principles give a background to the search for the specific factors that are the determinants or causes of particular geological situations. In complex nonideal systems it is the requirement for the identification of the greatest possible number of such specific factors that limits predictability. For example, observations of slipping in geological fault regions may give an approximate time interval during which an earthquake might occur. Yet, the actual disturbance, major or minor, that triggers the event may remain unrecognized and prevent a more precise timing of a catastrophe.

ATMOSPHERE, OCEANS, AND THE PREDICTABILITY OF WEATHER

The oceans and the atmosphere are the "softly flowing" inanimate robes wrapping the hard surface of the earth. Ocean tides are tangible reminders of the gravitational interactions between the masses distributed throughout the cosmos. Seen from far above, swirling clouds give the earth its seemingly unique and welcoming image. Oceans and atmosphere have changed on a long-range time scale and with large climatic and geological modifications occurring during the successive phases of geogenesis. On a short-term time scale, ocean-atmosphere interactions generate the major weather patterns that dominate our lives. These global interactions strikingly show the characteristics of nonideal systems. An example is the winds that carry moisture from the Indian Ocean into the Indian sub-continent, sustaining its agriculture and meeting the nutritional needs of its population. These winds are known as the monsoons. At irregular intervals the monsoons fail to come, and India's population suffers severe famine. The clue to these irregularities is not found within the meteorology of the Indian Ocean region itself but coincides with certain fluctuation, called *southern oscillations,* in rainfall, wind patterns, and oceanic currents in the tropical Pacific.

This region is dominated by two opposing meteorological conditions. In one of these, northeasterly and southeasterly tradewinds converge at the equator. At the same time, an oceanic current moves cold water northward along the west coast of South America, then westward along the equatorial region to form the south equatorial current. This cold current carries a plenitude of marine life and provides a rich fishing ground. However, the northeast and southeast tradewinds, coupled with the lower sea surface temperature generate arid conditions along the western portion of South America, in particular Peru, and at the same time, induce precipitation over the western Pacific. The exaggeration of this complex of meteorological factors has been named La Niña (the girl). At times its influence will extend as far north as North America, where a severe La Niña creates droughts, such as that experienced in 1982.

Normally, a transient balancing meteorological trend arises during the beginning of the calendar year: southeast tradewinds will bring an influx of warm ocean water along South America's west coast. During most years, this has only limited regional effects. Under exceptional conditions this meteorological trend intensifies. The cold ocean current of La Niña becomes displaced westerly from the coast and its characteristic rich marine life is replaced by different and sparse marine species adapted to the higher ocean temperatures. The resulting heavy precipitation in the adjacent South American continental region promotes a spurt of abundant agriculture. This meteorological condition has been named El Niño (the boy). Remarkably, just in the years of El Niño, with heavy precipitation and harvests of ample produce in western South America, an absence of monsoons has left India with drought and famine (Philander 1989; Walker and Bliss 1932).

What do the stories about La Niña and El Niño tell us about nonideal systems? The two systems are very delicately balanced. Even a slight weakening of the tradewinds diminishes the movement of the warm surface waters westward and thus the welling up of cold currents, and will cause a feedback, further weakening of the tradewinds, enhancing the El Niño beyond the normal limits (Bjerkness 1969). Therefore, the prediction of meteorological patterns in the Pacific depends on the recording of the intensities and directions of the tradewinds and of the temperatures on the surface and at several depths of the ocean at many locations. Several complementary air pressure systems over the Pacific Ocean seem to help maintain the normal variations of the meteorological patterns in this region. The disturbances that directly precede and give rise to El Niño have now been recognized, and computer projections predict its development. Just what and how the preconditions will appear cannot be predicted at present. It is noteworthy, therefore, that the catastrophic dimensions of the 1997 El Niño were not anticipated. We are dealing with an open, nonideal system in which any change in the state of the universe—solar radiation, bypassing of asteroids, irregularities in the subduction of the ocean floor—could generate a sufficient disturbance to start the (self-enhancing) build-up of El Niño.

The limits of predictability have also been clearly demonstrated for another class of major tropical weather systems, hurricanes. Here the necessary main condition for the initiation of a hurricane is a well-defined thermodynamic disequilibrium between the temperature of the ocean surface, the depth of a high temperature layer, and the moisture-saturation of the air above the ocean. These conditions are relatively widespread in tropical areas; similar imbalances also occur in the polar regions. Hurricane-type storms are relatively rare, and some small initial disturbance must occur to convert the potential of meteorological conditions into an active storm. Examples of triggering events are extensive systems of thunderstorms or air waves—pressure ridges—traveling across the breadth of the Atlantic. The frequency and effectiveness of triggering conditions is presently not predictable (Emanuel 1988), although the triggering events for smaller storms are often predictable, as are certain other tropical weather systems.

The relatively high predictability of extensive tropical weather systems is derived from large sets of identifiable factors that define the necessary, if not sufficient, conditions for the meteorological changes. In general, the predictability of local weather conditions is less dependable. During certain periods, local meteorological conditions become unstable enough for very small disturbances to initiate unpredicted events. This has become known as the butterfly effect—under unstable meteorological conditions the flapping of a butterfly wing could be thought to result in a tornado or at least a thunderstorm, via a concatenation of chance events begun by the energy imparted to the air by the butterfly. The main aim of prediction is then to identify unstable conditions that will enable the forecaster to estimate the probabilities for the actual occurrence of certain weather changes. In some cases, computer analysis of as many as 10,000 weather reports (each containing a variety of variables) from all over the world make such evaluations possible.

In ideal closed systems, one deals with fully defined relationships of a few variables on a single level of dimensions: electron-photon interactions in electromagnetism, molecular motion in statistical mechanics, interactions of macroscopic bodies in mechanics, and space-time in relativistic mechanics. Inanimate, nonideal systems still exhibit a relatively low level of complexity in their organization. For example, we distinguish between the levels of planetary systems, of galaxies, and of the universe as a whole; the subcontinent-continent earth surface organization; atmosphere-ocean subsystems, and the total terrestrial weather pattern. However, the number of factors determining the state of these open, nonideal, inanimate systems is too great to allow a full definition based only on the general principles derived from ideal systems. General principles are still of relevance but are taken for granted and the main effort is directed toward the identification of specific conditions that determine the particular event. For example, the principles of mechanics are of secondary relevance for understanding the assembly of the South American continent, and relativistic mechanics does not tell us where in the universe space-time will become deformed due to the collapse of a star. The uniqueness of the state of nonideal, inanimate systems is the result of particular interactions between the components of these systems in accordance with general principles (continental aggregation, thermal convection currents). The much higher complexity of nonideal living systems is now conceived as the result of highly specific interactions between the system components, without reference to general principles, described in more detail in the following section.

THE COMPLEXITY OF LIVING SYSTEMS

Living organisms offer favorable conditions for the study of highly complex systems with multiple levels of hierarchical organization. It is decisive that a variety of methodologies is available for a far-reaching analysis of living systems at the different levels of biological organization and the identification of their components. The possibility of such an analysis enhances the importance of studies of living systems for the understanding of complexity in general and for understanding the transition from complex, inanimate systems to sociopolitical systems in particular. In view of its possible importance, the complexity of living systems is more broadly discussed here in more detail than the complexity of the systems dealt with previously.

The emergence of biological complexity on earth depended on a coincidence of multiple elementary factors. The extremely hot radiation caused by the Big Bang explosion had to cool down to a temperature range that was sufficiently low for the conversion of radiation into matter and sufficiently high for the occurrence of cosmic turbulences of matter that produced the galaxies. Within the galaxies, planetary systems had to form, including stars like our sun, that in turn produced a fairly constant outflow of radiant energy over billions of years. The stability of the sun and its radiation depends on the ratio of electromagnetic and gravitational forces acting upon its matter. Early on, a shift in this ratio would have resulted in its inward collapse or in its expansive disintegration and the cessation of its radiation well before biological evolution had a chance to begin. To become a carrier of life, a planet had to revolve around the sun at a distance at which solar radiant energy would yield a life-sustaining temperature range on the planet's surface. Our planet had to be of the right size so that its gravitational field would hold together the atmosphere, the oceans, and the continents. And finally, conditions and materials available on the planet had to be appropriate for the synthesis of organic compounds such as amino acids, nucleotides, and possibly lipids for eventual assembly into functioning cells. Considering the low probability for the coincidence of all these factors, the evolution of life here appears to have been a process that was very rare, perhaps even unique.

The justification for the single heading of "life" for the diversity of biological phenomena was the recognition of some common characteristics: organization into cells as the smallest units maintaining the living state, reproduction, sensitivity and response to external stimuli, synthesis and degradation of cell constituents, metabolic energy production and use, developmental differentiation, genetic transmission of hereditary characteristics, and the evolution of successive living forms.

The studies of these characteristics initially followed separate lines of investigation and led to a compartmentalization of biology. The interaction of organisms with their environment became the main theme of ecology. Systematics and taxonomy were devoted to the elucidation of evolutionary relationships of species on the basis of morphological and biochemical similarities and differences and geographic distribution. The description of the macroscopic and microscopic structure of organisms and of their embryonic development became the subject areas of anatomy, histology, and embryology. The functional aspects of organisms, such as muscle contraction, nerve conduction and excitability, digestion, and growth and reproduction, were studied by physiologists. Human and vertebrate anatomy and physiology were closely associated with medicine. Initially, most of these studies suggested that life is unique not only in its origin but in its characteristics. There was never any doubt among biologists that the expanding universe, the shifting continents, and the interactions of oceans, lands, and atmosphere—the systems considered in the preceding sections—were specific instances of general physical states. In contrast, however, the properties of living matter seemed for a long time incompatible with what was known about physical systems, and suggested that there was some general force or power absent in nonliving matter that was driving the vital processes (vitalism). Even Claude Bernard, a leading physiologist of the nineteenth century, still invoked vitalism in the interpretation of life as indicated by the following quotation from one of his treatises (quoted in Loeb 1916).

> There is so to speak a preestablished design of each being and of each organ of such a kind that each phenomenon by itself depends upon the general forces of nature, but when taken in connection with the others it seems directed by some invisible guide on the road it follows and leads to the place it occupies. . . .

We admit that the life phenomena are attached to physicochemical manifestations, but it is true that the essential is not explained thereby; for no fortuitous coming together of physico-chemical phenomena constructs each organism after a plan and a fixed design (which are foreseen in advance) and arouses the admirable subordination and harmonious agreement of the acts of life. . . .

We can only know the material conditions and not the intimate nature of life phenomena. We have, therefore, only to deal with matter and not with the first causes or the vital force derived therefrom. These causes are inaccessible to us, and if we believe anything else we commit an error and become the dupes of metaphors and take figurative language as real.

Determinism can never be but physico-chemical determinism. The vital force and life belong to the metaphysical world (p. 3).

Up to the early twentieth century, vitalism was still invoked to explain how each one of the two separated halves of an early-stage sea urchin embryo could form complete sea urchin larvae. According to strictly mechanistic principles, further development of halves should not have been possible or should have led to formation of only halves of the complete larvae. This problem was solved much later. Meanwhile, the mechanistic interpretation of life gained importance.

Supporting the mechanistic approach was the conservation of energy and the production of heat and CO_2 in calculable quantities from biological "work," comparable to the energy production in combustion engines. This generated the idea that living organisms were merely physical systems of some type. The mathematical expression of physiological processes also seemed to demonstrate the relevance of physical concepts in understanding living systems. The effect of strictly physical-chemical conditions on biological processes such as fertilization and on responses to stimuli further strengthened the view that biological activities could be subsumed under the same concepts that also define inanimate systems (Loeb 1912, 1916). This led to several levels of interpretation of the organization of biological structure and function in terms of molecular processes, which provided both conceptual continuity and a clear demonstration of the complexity of the living state, discussed more fully in the following section.

THE MOLECULAR INTERPRETATION OF CELLULAR PROCESSES

Analysis of the molecular components of biological material began in the nineteenth century as an empirical subdiscipline of organic chemistry and was

aimed mainly at the preparation and characterization of small-molecular constituents of plants and animals. It began to emerge as a separate field with the
investigation of large-molecular biological substances that became known as
proteins and nucleic acids, and with research on the metabolic relationships of
the small molecular compounds (Olby 1990). For three reasons this new field,
named biochemistry or physiological chemistry, initially developed in isolation. First, organic chemists disdained work with still ill-defined material;
second, physiologists and biologists refused to see a relationship between the
macroscopic and molecular representations of the living state. One of the
distinguished physiologists of that time, J. B. S. Haldane, proclaimed:
"The new physiology is biological physiology—not biochemistry or bio-
physics. The attempts to analyze living organisms into physical and chemical
mechanisms is probably the most colossal failure in the whole history of modern science" (see Hopkins 1949, p. 181; see also Haldane 1931; Morgan 1990).

   As the third reason, biochemical investigations were carried out with biological material that was reduced to its simplest homogeneous state. The biochemist
would isolate and characterize the constituents of these tissue homogenates and
follow their transformation by metabolic enzymes and express the results quantitatively as numerical relationships. In contrast, for the biologist, form and
structure represented the main attributes of living matter, and maintenance of
the living material in its most complex form was one of the prerequisites for
meaningful studies. It became a cliché to argue that the whole organism is so
much more than the sum of its parts that the analysis of the homogenates of the
biochemist was of no relevance for the understanding of the living state. The
need for a resolution of this impasse was cogently advocated by F. G. Hopkins, a
leading personality in the development of biochemistry and molecular biology.
Hopkins pointed out that the biochemist's aim is not just the characterization of
isolated molecular components of the living organism, and that even the understanding of a "whole" molecule depends on the understanding of the interactions or relationships of its respective parts. In the same manner the biochemist
must be concerned with the relationship of the molecular components in
forming cell structures and in producing the functional changes of the living
state. Hopkins referred here to Whitehead's *Philosophy of Organism* in dealing
with the whole-part relationship in general. In his own terms, Hopkins adds the
following comments on this decisive point (Hopkins 1949).

> Biology has always had to relate the behavior of the objects it studies to the properties
> of highly complex systems considered as a whole; whereas the exact sciences long

progressed owing to the fact that their material allowed them to reduce most or many of the problems to such simple terms that the essential significance of the data remained independent of underlying mechanisms which need visualization. On such lines thermodynamics developed with the complete independence of mechanistic considerations. This independence is the essence of the power and attractiveness of its methods, but that power is strangely lessened with phenomena which remain under conditions of which the very essence is complexity. Complexity is the essence of the material systems which manifest the phenomena of life; it cannot be eliminated from the only kind of picture of such systems which is real (p.184).

To Hopkins, atoms, molecules, or organisms are units that exist as wholes by virtue of the relatedness of their parts. The understanding of any such whole depends on the understanding of the mechanism of the interaction of its parts. Hopkins says:

The atom we now find is neither a pure abstraction nor a homogeneous unit, but a mechanism or organism in which the properties of the whole depend not solely on the nature of its parts but also on the peculiar relatedness of its parts, while an organic molecule is an actuality and no convention and, again a mechanism or organism in which the properties of the whole depend on the relatedness of its parts as well as on their nature.

Hopkins continues:

We can only properly define the whole in terms of its own activities. Neither the atom nor the molecule is a static system. The behavior of each depends not solely upon its structure, but also upon the nature of events which occur in that structure. We have to ascertain in each case how the atom or molecule functions as a whole just as we have to in the case of what we are more accustomed to call an organism. In all cases, structure, internal events, and functions must be thought of as inseparable if we are to grasp the essential nature of such systems; but as everyone will admit (in the case of the atoms or molecules at least) we could hardly obtain that grasp if we had not knowledge of the parts to subserve our mental synthesis (Hopkins 1949, p. 186).

Hopkins's appraisal and expectations were fully borne out by later developments in biochemical investigations that turned from the isolation and characterization of individual products of cell metabolism to the exploration of the sequences of metabolic reactions. In the presence of oxygen, metabolic pathways of animal cells were found to convert carbohydrates, lipids, fats, and amino acids into carbon dioxide. The establishment of the main pathways of metabolic energy conversions—such as oxidations-reductions, phosphorylations, their control by diverse feedback mechanisms (directly for example by insulin and other hormones)—gave an early picture of the molecular basis of biological complexity (Hill 1950; see Morowitz 1968; Harold 1986). Linked to enzymatic oxidations are other molecular reactions, which in turn activate further cell processes such as muscle contraction, secretion, nerve conduction, and the synthesis of cell constituents.

Thus, beginning with the isolation and characterization of small molecular tissue components, subsequent coordinated molecular and electron microscopic studies showed that enzymes involved in metabolic oxidations and the receptors of cell activators are intrinsic parts of cell membranes or of the contractile mechanism. Further, the transmission of genetic information and cell reproduction involved synthesis of molecular components that depend on the association of enzymatically active macromolecules with specific cell structures. Molecular biologists began to think of the cell organization as a whole, which maintains at the same time a conceptual continuity between its parts. In the course of these developments, biochemistry was transformed into a molecular biology that established the link between the simplifying generalities of physics and the specificities of complex biological systems. It also generated a host of biotechnological applications. To give the concept of biological complexity and conceptual continuity a concrete meaning, the molecular elements and their highly specific interactions are discussed below in terms of two examples, the contractile system of muscle and gene expression.

THE CONTRACTILE SYSTEM OF MUSCLE

Muscle tissue contributes a large part of the bulk of higher organisms, and can be used as a model for clarifying the role of molecular analyses used to establish a conceptual continuity in complex living systems (Herrmann 1989).

Skeletal muscle is highly organized; the number of well-defined levels of tissue and cellular structure may be greater than in other tissues; at least, it is more obvious (figs. 4.1 and 4.2). Individual muscles are sheathed by a collagenous membrane (epimysium) from which septa extend into the muscle interior (perimysium) and divide the muscle into muscle bundles or fascicles a few millimeters in diameter. The fascicles are subdivided into 20–40 muscle fibers of 20–80 um diameter, each fiber covered by a fine collagenous membrane, the endomysium. Each muscle fiber has an electrically polarized surface membrane and comprises a large number of 1–2 um diameter myofibrils. Under the light microscope the fibrils show alternating denser, birefringent, anisotropic segments and less dense, isotropic segments, known as A bands and

I bands, respectively, that produce the characteristic cross-striation of skeletal muscle. The cross-striation is also seen in the smaller muscle fibers. In addition to the A and I bands, a less dense region is seen in the center of the A bands (the H region). A dense structure, the Z disc, traverses the I band.

About twenty specific proteins have been identified as responsible for this high level of visible organization. Two main types of proteins, myosin and actin, make up separate but partially overlapping and closely adjacent filaments that are the basic elements of both muscle structure and function. Extensions of the myosin molecules form crossbridges that link the myosin and actin molecules. In action, change in the conformation of the crossbridges effects the sliding of the myosin and actin filaments past each other, resulting in contraction of the muscle. A pivotal experiment demonstrated that the conformational change in the myosin crossbridge is due to the bonding of adenosine triphosphate molecules transferring the energy generated in the course of carbohydrate metabolism to the contractile mechanism and thereby to cell and tissue function in general.

The interaction between the myosin crossbridges and actin is controlled by an actin-associated complex of four additional proteins, tropomyosin and three forms of troponin. In resting muscle, this complex prevents the myosin-actin interaction. When stimulated by a nerve signal, calcium ions are released from internal storage, the ions bind to the control complex, inducing a change in conformation that permits myosin-actin interaction and the sliding of the filaments that is the macromolecular basis of contraction.[2]

Through this macromolecular interpretation, muscle cytology and physiology converge into a consistent common framework of molecular concepts. In this sense, the past five decades have generated a conceptual continuity in our thinking about muscle structure and function that may be an example for the structure-function unification of other cell processes as well. It is most relevant that in contrast to the ideal systems in which very general elements create continuity, in nonideal systems this is achieved by highly specific elements. The following description of gene expression and replication shows that this type of unification has become the prevalent approach toward understanding complex living systems at the cellular level.

GENE EXPRESSION AS AN EXAMPLE OF BIOLOGICAL COMPLEXITY

Gene expression is now widely seen as a well-established relationship with genetic "information" stored in large DNA molecules being converted into the

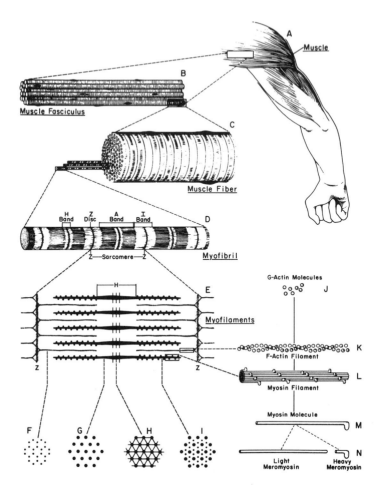

Figure 4.1 Levels of Organization of the Contractile Muscle System.
(A) Intact muscle. (B) Bundle of multinucleate muscle cells (myotubes). (C) Enlargement of a single myotube showing individual myofibrils. (D) Enlargement of a myofibril showing the pattern of banded sarcomeres that can be seen with the electron microscope. (E) Enlargement of a single sarcomere showing the overlapping thick and thin filaments of which it is composed. The thick filaments are myosin and the thin filaments actin. (F) Cross section through the thin actin filaments. (G) Cross section through the thick myosin filaments. (H) Section through the midpoint of the A band, showing cross connections among the thick myosin filaments. (I) Cross section of the overlapping region of the thick filament. (J) Globular subunits of actin. (K) Actin monomers polymerized into an actin filament. (L) Thick myosin filament with globular projections. (M) Individual myosin molecule, rod-shaped, with globular projection at one end. (N) Myosin molecule cleaved by brief exposure to trypsin, forming rod-shaped light meromyosin and globular heavy meromyosin, which is the part of the molecule that interacts with actin. (Reprinted by permission from D. W. Fawcett and W. B. Saunders, 1986.)

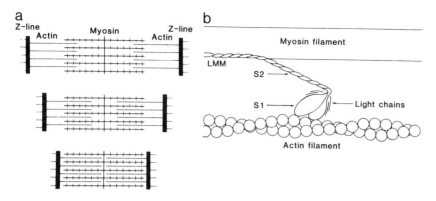

Figure 4.2 Interaction Between the Main Muscle Proteins During Contraction. (a) Diagram of a single segment (sarcomere) of a muscle fiber as it shortens. The shortening of thousands of segments in series along the muscle fiber causes the whole muscle to shorten. The short lines perpendicular to the myosin filament represent the myosin cross-bridges. (b) Diagram of a myosin molecule bound to an actin filament in the absence of adenosine triphosphate. Both the myosin and actin filaments are polymeric structures. Each circle in the actin filaments represents an actin monomer. Only one myosin molecule is shown; in vertebrate skeletal there is one myosin cross-bridge present for every two actin monomers. During contraction, the actin filament moves to the left relative to the myosin filament. (Reprinted by permission from E. Eisenberg and T. L. Hill and the American Association for the Advancement of Science, 1985.)

"properties" of living organisms.[3] Actually, in its crude form this relationship merely indicates a correlation between two separate concepts. One of these is DNA and the other is some property such as the eye color of the fruitfly, the coat pattern of a mouse, or a big lower lip in a member of the Habsburg dynasty. Apart from such correlations, the related items—that is, DNA and the cellular and organismic properties—are quite different things and at that stage there is no conceptual continuity that leads from DNA to the respective genetically determined characteristic. Advances in the understanding of the series of specific steps by which genes are expressed are not only establishing conceptual continuity between genes and cell characteristics but are also indicating the meaning of biological complexity.

Isolated DNA is a thin molecule only a few thousands of a millimeter in diameter. In higher organisms it reaches a length of several centimeters in its longest form. In turn, it is made up of linked units of nucleotide molecules (each about molecular weight 300). Each nucleotide includes either an organic purine or pyrimidine base: a sugar (deoxyribose corresponding to the D in

DNA); and finally, a phosphate group attached to the sugar components of two adjacent nucleotides. Several hundreds of these nucleotides are linked by their sugar-phosphate moieties into the long DNA filaments. In a somatic chromosome, two DNA filaments are held together by weak hydrogen bonds between pairs of purine and pyrimidine to form the DNA double helix.

In unicellular prokaryotes such as bacteria, the DNA is loosely folded, readily accessible for interaction with molecular regulatory factors. In multicellular organisms the DNA is sequestered in the cell nucleus, separated from the cytoplasm by an envelope that includes pores for transport of large molecules to and from the inner space. The envelope, the DNA, and a matrix of fibrous molecules constitute the cell nucleus. In the nucleus the DNA is bound to clusters of basic proteins, the histones. The histone-DNA complexes are highly compacted by several phases of coiling and are known in this state as chromosomes.

The chromosomal DNA-protein complex can be readily seen under the light microscope and has been the object of intensive investigation for more than a century. Chromosomes show a banded structure; defects in certain bands were associated with specific abnormalities in fully developed organisms, suggesting the location of specific "controlling" genes in the chromosomes. Later on, advances in technology permitted the selection of the gene location in the chromosome to the DNA molecule itself. To find out how the genetic information, stored in the DNA, is converted into cell characteristics became the main task.

Structural and functional cell characteristics, such as the contractile apparatus of muscle, the neural units of the nervous system, the digestive functions of the intestine, the keratinization of the skin, are all the result of the synthesis of cell-specific proteins. The genetic information for synthesis of proteins consists in triplets of the DNA nucleotides, with each triplet being the molecular "code word" for an amino acid, the small-molecular units from which large protein molecules are assembled. A protein molecule is made of up to a hundred or more amino acids, and a corresponding number and sequence of DNA triplets provide the information for synthesis of a whole protein molecule. In addition to the triplets serving as the code for a certain type of protein, each gene or group of genes includes DNA sections that are binding sites for control proteins that activate or deactivate the initial step in the use of genetic coding. When activated by a regulatory protein, the respective gene sequence of DNA is converted by a protein enzyme in the cell nucleus into another form of nucleic acid that contains ribose instead of deoxyribose as the sugar moiety, called ribonucleic acid or messenger RNA. The transfer of genetic information from DNA to RNA is known as transcription.

RNA molecules are released into the cytoplasm and there attach to structural units, the ribosomes, where the genetic coding patterns is converted into specific proteins. A further, small molecular RNA species binds the amino acid components of the prospective protein and aligns them in accordance with the genetic information in the messenger RNA. The aligned amino acids are then enzymatically linked into amino acid chains that assume the three-dimensional conformation of protein molecules by association with "chaperone" proteins. This step is known as translation.

Finally, the protein end products of translational synthesis are transported through systems of vesicular and filamentous cell organelles to the right sites in the cells or are released from the cell for use, namely, as hormone or immunoprotein. The conversion of genetic coding through the long chain of specific intermediate steps can be modified at each point. In the cell nucleus, for example, transcription of specific genes is activated by agents such as epidermal or nerve growth factors. In the cytoplasm, adjustments, for example, in ionic composition, regulated by ion transport units in the cell surface layer, can increase or decrease the efficiency of the translational process.

Individual groups of genes can be activated by specific factors without activation of other genes. This leads to the appearance during development of differentiating cell types that, for example, may produce muscle proteins for contraction, neural structures for transmission of nerve signals, or bone material to support the weight of the body. The main point in this detailed description of gene expression is to show how the simple and general but conceptually discontinuous relationship of gene (or DNA) and organismal characteristic is replaced by a complex sequence of highly specific interactions of macromolecules that establish the conceptual continuity of gene expression.

THE MIRAGE OF SIMPLIFICATION AND THE VISION
OF COMPLEXITY IN BIOLOGICAL REALITY

Inherent in the scientific thought of past centuries has been the resolve to create a representation of reality that is free of complexity. Examination and discussion of complexity as a noteworthy concept in its own right was to be avoided. Phenomena that could not be reduced to the simplicity of ideal systems were to be disregarded. Ignoring complexity seemed quite plausible, as long as some mathematical formalization could be regarded as the only necessary and sufficient requirement for understanding of nonideal systems. This tendency became particularly frequent in dealing with biological systems. Physiology texts published at the turn of the century could boast that their mathematical correlations of physiological parameters gave physiology a status comparable to that of physics. We now see that these "laws" of physiology correspond only to the correlations of classical physics (such as the "laws" of optics or electromagnetism) of the nineteenth century. It was an article of faith that "The ideal axiom at the basis of all causality can only be stated in terms of the mathematical concept of functions. All sciences are special chapters of mechanics" (Meyer 1934). Since mathematical formalizations of reality were widely regarded as the most satisfactory representations—even of complex systems—some limitations of this approach should be pointed out here.

In the representation of nonideal biological systems the mathematical formalization was carried to an extreme in a formidable treatise entitled "Growth and Form" by D'Arcy Wentworth Thompson (1916, 1942). It was greeted with substantial accolades and its author was heralded as "an aristocrat of learning whose intellectual endowments are not likely ever again to be combined within one man" (Medawar 1967). Thompson was impatient toward what seemed to him the purely descriptive homologies and analogies used by comparative anatomists, as the major trend in the nineteenth-century evolutionary interpretation of organismic form. Instead, he fervently defended his conviction that any kind of mathematical formalization leads to the understanding of biological phenomena. In support of this, he showed in numerous examples that growth can be represented by simple, mathematically defined curves and that different shapes—for example, of skulls or bodies—could be conceived as transformations of geometric coordinate systems, and that some features of living organisms—namely, spicules—could be regarded as the result of general adsorption and crystallization processes. Thompson actually spoke about a general morphogenetic "force" defined by his growth equations and geometric transformations.

Thompson's approach is specifically mentioned here because his mathematical formalization of growth was not just a passing fad but was taken up by others and developed further, with increasing sophistication of the mathematics involved. Even though the field came to be called "Principles and Theory of Growth" (Bertalanffy 1960), reservations concerning Thompson's treatment also hold for the later, more advanced mathematical interpretations. Even in this more elaborate form we are dealing with correlations that give useful descriptive baselines for growth phenomena. They do not, however, give information about the multiple factors that come into play at the different phases of the growth processes, nor do they provide a basis for conceptual continuity. Similarly, most nineteenth century physiological "laws" are of a descriptive nature.

We now know that there is no general "force" that produces the described changes. Even in unicellular prokaryotes cell multiplication requires twenty-one different proteins for the replication of the DNA molecule alone, and numerous additional factors are involved in the unfolding and refolding of the DNA molecule and in regulating the replication process itself. Also, the bulk of the body of higher vertebrates, made up of bones, skin, and muscle, is the result of cell replication and nonreplicative increases in cell and tissue mass. In plants, growth may be due to an enormous increase in cell size with formation of cell vacuoles containing mostly water. In general, during the initial phase of development in plants and animals, growth depends on cell replication. With the onset of cell differentiation, replication slows down or ceases, and the synthesis of cell-specific products begins. Since the publication of the first edition of Thompson's treatise in 1916, it has become evident that biological phenomena, such as growth, are regulated by several highly specific hormonal, neural, and other factors. After these factors have been identified, mathematical modeling can help in deciding which groups of factors control and regulate the different phases of the overall growth process. A general mathematical formalization such as Thompson's, without reference to specific process components, misleads by giving the impression of an understanding of the underlying complex mechanism of biological phenomenon. Yet, simulation as developed in systems theory or mathematical modeling of specific attributes of living organisms is in much use.

A further example of a mathematical interpretation of biological processes is the use of chaos theory in predicting the onset of lethal arrhythmias of heart beat in humans. The mathematical simplification "of classical nonlinear behavior is readily explained at a molecular level by the physical relation between actin and myosin filaments" (Denton et al. 1990, p. 1421). This shows that a mathematical formalization of nonlinear forms of change can be of high predictive value, but it requires the identification of both specific and complex interactions of the components of the system to provide conceptual continuity in understanding the mathematically defined phenomena.

Another attempt to simplify the representation of biological systems and to avoid direct confrontation with biological complexity led to a search for new physical principles, as general as those of mechanics, but applicable to the interpretation of life, by a group of physicists during the middle of this century. The credo of this group was summarized by Max Delbrück who was the leading personality in this avant-garde, and who attempted to carry physical thinking into biology:

It may turn out that certain features of the living cell, including perhaps even replication, stand in a mutually exclusive relationship to the strict application of quantum mechanics (meaning here interpretation by molecular concepts in general), and that a new conceptual language has to be developed to embrace this situation. The limitation in the applicability of present day physics may then prove to be not the dead end of our search, but the open door to the admission of fresh views of the matter. Just as we find features of the atom, its stability, for instance, which are not reducible to mechanics, we may find features of the living cell which are not reducible to atomic physics but whose appearance stands in a complementary relationship to those of atomic physics.

This idea, which is due to Bohr, puts the relation between physics and biology on a new footing. Instead of aiming from the molecular physics end at the whole of the phenomena exhibited by the living cell, we now expect to find natural limits to this approach, and thereby implicitly new virgin territories on which laws may hold which involve new concepts and which are only loosely related to those of physics, by virtue of the fact that they apply to phenomena whose appearance is conditioned on not making observations of the type needed for a consistent interpretation in terms of atomic physics (Delbrück 1949, pp. 188–189; see also Cairns, Stent, and Watson 1966).

Delbrück became the prophet of this faith, and Schrödinger, one of the founders of quantum mechanics, wrote its bible in his essay *What Is Life* (1944, 1967).

One would have expected that this approach was aiming at a purely speculative interpretation of life phenomena. However, the members of the Delbrück group were experimentalists who supported whatever conceptual schemes might be postulated with new experimental data. Their biological model had to have at least the appearance of a basic simplicity, something like the hydrogen atom of classical physics. With reproduction, the most puzzling characteristic of living matter, the then newly discovered replication of viral phage particles in bacteria was adopted as a promising system. Ironically, as experimentation progressed, the idea of an abstract, new representation of the living state became a mirage that receded beyond the horizon. The attempt to create a new class of concepts defining ideal biological systems and the desire to avoid the stigma of biological complexity failed. Instead, the experimental work with phage became one of the foundations of molecular biology and, in the extended perspective, one of the approaches that led eventually to biotechnology (Cairns, Stent, and Watson 1966). With another dream of simplifying abstraction of biological systems ended, complexity began to acquire respectability after centuries of denial. The relevance of recent attempts (Kauffman 1993) to develop a more abstract theoretical foundation of life processes, including development and evolution, remains to be evaluated. Broad philosophical interpretations of biology are provided by Sattler (1986) and Rosenberg (1985). The problematics of theoretical abstractions for the understanding of complex systems has been presented by Strohman (1997).

REEVALUATING IDEAL AND NONIDEAL SYSTEMS

With the survey of examples from several scientific fields, we can ask in what sense ideal and nonideal systems represent simple and complex reality, respectively. Certain characteristics of ideal systems are here recapitulated as a basis for comparison. Kepler's laws provided an initial mathematical representation for planetary motion. However, the masses and the motions of the planets were still conceptually separated, and the basis for the demonstrated mathematical relationship was unknown. Kepler's laws, although mathematically formalized, are therefore merely descriptive, phenomenological laws (Cartwright 1989) that seem to indicate a precise invariance in occurrence without any conceptual link between the related magnitudes. This deficiency was tentatively eliminated by Newton's introduction of "force" as the link between mass and motion, which allowed an integration of three general variables in one mathematical equation that expresses a fundamental law and is the basis of a theory proposed as a unifying principle. It is the abstractness and generality of the related variables that results in the simplification and generality of such a mathematically defined relationship.

In Newton's mechanics only a few variables determine the state of the planetary system, and for this reason it can be called a fully defined system; the planetary system is an abstraction that excludes interfering factors such as friction, and it disregards specific mass properties such as hardness, crystallinity, and conductance. Such fully defined abstractions represent ideal systems that are removed from the context of the universe (closed systems), where asteroids and comets could interfere with Kepler's laws, or changes in the sun could affect the conditions of the planets. Actually, recent computations indicate deviations in the planetary behavior described in terms of chaos theory, which show that the Newtonian universe is not a fully defined ideal system (Sussman and Wisdom 1988, 1992; Touma and Wisdom 1993).

Mechanistic generalizations do not inform us about the nonideal specifics of planetary systems, their composition by varying numbers of planets and moons, the distribution of stars and planets in the respective galaxies, or of the galaxies in the universe. Many factors besides gravitational force determine the specific dynamics of terrestrial landmasses, of long-range changes in climate, of short-range changes in the earth's atmosphere, and of the evolutionary diversity of living forms on the earth. Establishment of conceptual continuity in the representation of these systems requires inclusion of the largest possible number of related variables. We need to know the origin and assembly of all geological terrains and the roles of subduction in their movements and their thrusts in the formation of the South American continent; similarly we need to know the locations and specificities of regulatory factors controlling muscle contraction or gene activation and expression. The minimizing of variables in ideal systems generates the simplicity of closed systems. The maximizing of variables in nonideal systems is required to adequately represent the complexity of open systems and to achieve ultimately conceptual continuity.

The abstractions of ideal systems are fundamental to understanding nonideal systems; the mechanics of continental drift; the role of quantum mechanics in the phenomena of molecular biology, but they do not directly contribute to our understandings of the specificities of nonideal phenomena. The relationship of ideal abstractions to nonideal specificities is here again compared metaphorically to the relation of grammar and literary expression. Evidently, the ground rules of grammar have to be observed in a sonnet, but for the artistic expressiveness of a poem, the emotional drive in a drama or in a novel, the grammatical ground rules are only of secondary relevance and a multitude of other factors yield aesthetic satisfaction (Campbell 1982).

The understanding of ideal systems of physics requires the elimination of specifics and depends on establishment of conceptual continuity between general aspects of the systems investigated. In contrast, understanding of nonideal reality requires establishment of conceptual continuity between the specific aspects of inanimate matter, the universe, the dynamics of landmasses, the oceans, the atmosphere, and the structure and function of living cells. Classical physics implies that forms of reality can be represented by simplification and abstract generalization. Alternatively, the analysis of nonideal systems suggests that perhaps the bulk of everyday experience is more compatible with a complex reality. This may be the main message that has emerged from the preceding discussion. The relevance of this message for the understanding of social systems is explored in the following chapters.

Chapter 5 Conceptual
Continuity and the High-Level
Complexity of Democracy

Reality is represented in this book as a spectrum of increasingly complex systems. At one end of the spectrum are the ideal systems of physics that establish constant relationships between the most general properties of matter and that are fully defined by only one or a few variables (see Chapters 2 and 3). On the next higher level, the nonideal systems of physics are described by the same general relationships but require the identification of specific subsidiary factors, such as the friction of the moving parts of an engine, for their full definition. A further increase in complexity, like that in biological systems, is still compatible with the laws and relationships that hold for fully defined systems. At that level of complexity, however, the abstract, general properties of matter (such as mass or radiation) are of little significance and the identification of a great number of specific elements or variables and the establishment of conceptual continuity between them is the primary concern here (see Chapter 4).

So far, the levels of complexity have been defined relative to the ideal systems of physics: at crucial points in Western history (the Enlightenment, Marxism, Classical Economics) even sociopolitical ideas have been represented as if they were basically similar to the constructs of physics or biology. In actuality, the interpretations of physico-biological and sociopolitical complexity differ fundamentally.[1] The latter is generally understood: "To refer to such things as the size of a society, the number and distinctness of its parts, the variety of specialized social roles that it incorporates, the number of distinct social personalities present, and the variety of mechanisms for organizing these into a coherent, functioning whole" (Tainter 1988, p. 23).

As a concrete example, the demography of successive periods of the history of Egypt was defined by certain system variables: A progressive build-up of a metastable sociopolitical hierarchy, the presence or absence of strong leadership, ecological conditions (in particular the water level of the Nile) and foreign intervention. According to this interpretation, "civilizations behave as adaptive systems, becoming unstable when a top-heavy bureaucracy makes excessive demands on the productive sector; breakdowns result from chance concatenation of mutually reinforcing processes not from senility or decadence" (Butzer 1980, p. 517). What is regarded as "chance concatenation" suggests the existence of additional variables indicating the complexity of such systems.

One of the main variables is the relationship of the individual to society. The presence of this variable suggests three main phases in the development of Western society. During the first phase, tens-of-thousands years of prehistory, the individual members of small groups of hunters-gatherers made distinctive contributions to the maintenance of the whole. The demands of mere survival of the group as a whole were so great that each member had to give nearly its full energy to the common cause with little consideration of the individual self. In hunting for big prey, in encounters with rival groups, and in healing rites and other ceremonials, some members of the group assumed, at least temporarily, somewhat exalted positions as leaders. Presumably such leading personalities remained cooperatively integrated into the group, precluding a ruler-ruled relationship. During this extended period, nearly ten times longer than the recorded history of the previous 6,000 years, the social status of the individual as a member of the group, not as a separate independent individual self, remained unchanged. Restrictions of individuality were imposed by the exigencies in the collective maintenance of the group and not by group members who assumed positions with dominance.

The individual-to-group balance changed dramatically with the development of the second phase, that of large-scale sociopolitical organizations, the empires of the ancient history. With the exception of the citizens of the small city-states of Greece and, later on, of the Italian Renaissance, the majority of people in Western culture existed for millennia in a two-class system of secular or clerical servitude. In this simple relationship, wealthy rulers who held all power ruled the subjects, who eked out their lives on a subsistence level. Peasants and laborers were essential for the maintenance of the ruling minority, but had no opportunity for establishing a sociopolitical identity for themselves. The balanced duality of being a part of the social structure as well as a whole individual did not exist for the majority.

Only during the third phase, the utopias of the seventeenth and eighteenth centuries, the individual emerged from a state of total subordination as citizen in his own rights and as active member of society. During this period Hobbes, Locke, the French Philosophes, and even Rousseau envisaged direct, simplified relationships between individual and society, with the recognition of areas of common interest, regarded here as preliminary forms of sociopolitical conceptual continuity (see Chapter 2).

In a parallel development, Montesquieu, Burke, and the founders of the American Constitution opted for the complexity of administratively separate, indirect forms of representation and multiple interacting public and private agencies, opening approaches to conceptual continuity as the balance between private and public, and between individual and society.

During the past two centuries the sociopolitical reality of the Western world became polarized into two competing trends. Following one of them, the individual became submerged in the one-sided simplifications of nationalistic and communistic ideologies. With the defeat of German nationalism, and the breakdown of the Soviet Union, these ideologies were thought to have passed their apogee (see Chapter 6).[2] They still continue, however, in the forms of ethnic extremism and religious fundamentalism in the Balkans and in many other places around the world, with communism still ruling the population of mainland China, North Korea and some Eastern European countries. Nationalism, communism, and fundamentalism all show the deep-seated craving of human nature for certainty and a simplification of existence that eases the burden of social responsibility.

In the other trend, the sociopolitical complexity of democracy was more widely adopted among Western societies. Democracy is a sociopolitical organization of high complexity for several reasons. Most fundamentally, with the emergence of democracy the individual faces, as an everyday experience, a dichotomy, emphasized by Keane in his introduction to Norberto Bobbio's *Democracy and Dictatorship:* "The human species is subdivided for the first time into social classes; individual's legal status is divorced from their socio-economic role within civic society; each individual is sundered into both private egoist and public spirited citizen; and civil society, the realm of private needs and interests, waged labor and private right, is emancipated from political control and becomes the basis and presupposition of the state" (1989, p. xxvi).

In a standard definition, democracy carries the labels of liberty, equality, and fraternity. It is that form of government which protects the basic rights of its citizens: freedom of speech and association and pursuit of happiness (meaning essentially the right to acquire and hold property). The citizens of a democracy elect those who carry out the responsibilities of governance, primarily the protection of their rights in accordance with a written constitution. The citizens expect that a democratic society will generate living conditions that seem adequate to all (see also Keane 1993).

The democratic emancipation of the individual opened a broad range of options for the realization of different forms of human existence, and established conditions for the direct or indirect participation of the citizen in the conduct of governance. The individual is expected to grasp the separate aspects of private and public existence, and to make choices between alternatives that cannot always be easily reconciled. Indeed, the individual is not to be sundered by a dichotomy of separate private vs. public compartments, but is expected to unite the two aspects of his or her sociopolitical reality as a coherent duality (conceptual continuity). In this sense the making of decisions is no longer a matter of the simple "either-or" but rather a matter of the more demanding "to what extent the one or the other" or "how much of each shall be implemented." In a democracy, the individual is made aware of the multitude of apparent alternatives to be considered in making decisions by increasing education and by the burgeoning news media. Again, as Keane (1989, p. xxiv) points out, it is therefore important to realize:

> Democratic procedures are superior to all other types of decision-making not because they guarantee both a consensus and "good" decisions, but because they provide citizens who are affected by certain decisions with the possibility of reconsidering their judgments about the quality and unintended consequences of these decisions. Democratic procedures increase the "*flexibility*" and "*reversibility*" of decision-making. They encourage incremental learning and trial-and-error modification . . . and that is why they are best suited to the task of publicly monitoring and controlling complex and tightly coupled "high risk" organizations whose failure can have catastrophic ecological and social consequences.

The flexibility and reversibility in decision-making and the general responsiveness of democratic society is to be ensured by its high degree of social organization and by the alertness and active participation of its citizens. The multiple controls needed for the adjustment of the private-to-public balance and of other opposites are the factors giving modern democratic society the complexity absent in the democracies of ancient Greece and the Italian city-states.[3] These are the adjustments that imply establishment of sociopolitical conceptual continuity. Therefore, it is necessary to define the nature of this complexity in more concrete detail to counteract the ingrained urge of the common man to reach for the simplistic formula, when reality demands complex adjustments to solve sociopolitical problems. Sociopolitical complexity is exemplified here by following its development in two democratic societies: the adversarial-competitive, majoritarian democracy of the United States, and the consensual democracy of Switzerland. The development of democracy can be seen as the quest for conceptual continuity between the basic components of the sociopolitical system. Therefore, the development of democratic organization and of sociopolitical conceptual continuity are regarded as parallel processes.

THE TRANSITION FROM SIMPLE TO COMPLEX AMERICAN SOCIOPOLITICAL REALITY

The difference between low-level and high-level complexities of society and its relationship to conceptual continuity is clearly exemplified by comparing the American colonies before and after their independence (see Harrington 1989; Nettles 1989). The colonists in the original Puritan Commonwealth were governed by a covenant in accordance with the laws prescribed by God in the Christian Protestant tradition. The administration of the colony was in the hands of elected magistrates, supervised by an oligarchy of elders and an elected governor. The act of voting gave the common man an opportunity for self-assertion as an expression of individualism. The actual relationship between the individual members of the colony, the magistrates, and the community as a whole had been set forth by John Winthrop, elected colonial governor from 1630–1649 (Dolbeare 1981, p. 26). Winthrop asserted that the magistrates had to distinguish between natural and civil liberty. Human beings in the state of natural liberty tend toward evil, and they will reject any authority that attempts to compel their existence into the pursuit of the common good. In this natural liberty Winthrop saw the wild beast in human beings that make them grow more evil and "in time to be worse than that wild brute beasts which all the ordinances of God are bent against, to restrain and subdue it." The other kind of liberty called by Winthrop "civil," "federal," or "moral" is the basis for "the covenant between God and Man in the moral law, and the political covenants and constitutions amongst men themselves. This liberty is the proper end and object of authority and cannot subsist without it and it is the liberty to that only which is good, just, and honest."

Winthrop held the following admonition before the community: "If you stand for your natural, corrupt liberties and will do what is good in your own eyes you will not endure the least weight of authority, but will murmur, and oppose, and be always striving to shake of the yoke; but if you will be satisfied to enjoy such civil and lawful liberties, such as Christ allows you, then you will quietly and cheerfully submit unto that authority which is set over you, in all administrations of it for your good" (Dolbeare 1981, II: 28).

But diversity in religious commitment was definitely discouraged, and even financial transactions were rigorously controlled. Prices for food and ordinary commodities and interest charges on loans above officially set levels were prohibited. Failure to comply was penalized with admonition, banishment, or even excommunication. Ostensibly, the commonwealth of Winthrop's time had a minimal degree of administrative structure with the Governor, the elders, and the magistrates safeguarding the implementation of laws prescribed by conservative, religious tradition. Where individual distinction existed in specialized craftsmen, ministers, or teachers, their roles were strictly communal and not individual; there was still only minimal duality in civic existence. The denial of a State of Nature, and hence of Natural Rights, prevented the emergence of liberal individuality in a community that for a short time represented American sociopolitical reality in both its most conservative and simplest form. Sociopolitical conceptual continuity, based on identification of areas of equal interest to both individual and society were still rudimentary.

This group ethic was soon superseded by a variety of alternatives. Thus, in Connecticut settlements founded in 1636 by Thomas Hooker, more extensive control of governance by individual, free people was instituted. Roger Williams established communities in Rhode Island (1636; charter 1644) that implemented full religious freedom and individual rights of conscience and free speech. "Sovereignty lay with the people; they created the state to serve their needs and preserve their individual rights" (Dolbeare 1981, p. 16). The church was completely separated from the state and was deprived of any authority in civil governance. These developments, along with modifications of the original Commonwealth governance represented a wide range of individual-society opposites, foreshadowing the effects of the writings of Hobbes, Harrington, and Locke, to be published during the second half of the seventeenth century.

> Instead of starting with the premise of an organic whole that assigned stations and obligations to people and demanded their loyalty and obedience, people would in time think of themselves as independent individuals with rights that came before the society—which was after all, only a body of many similar individuals (Dolbeare 1981, p. 14).

Individuality began to assert itself in the next developments. Still under the governorship of John Winthrop, a group of colonists introduced a Charter of the Massachusetts Bay Company, and the Reverend Nathaniel Ward was asked to draw up a set of laws that became known as the Body of Liberties (1641). These laws redefined the state of the individual within this Commonwealth with a list of the basic rights of individual existence that shall not be infringed upon or taken away without due warrant by the General Court, which acts as the main legislative and executive body of the government. It is not clear at this point just how far the individual could take his or her liberty, nor to what extent freedom of speech or assembly would be tolerated. Although the Body of Liberties refers to liberties of the individual, it is not an unreserved endorsement of individual freedom. Rather, its main purpose was a statement of independence of legislative decision-making as a matter of the colony itself, independently of the English Parliament.

Halfway between the Declaration of Independence and the drafting of the Constitution of the United States, John Wise, who served as minister to several parishes in New England (1652–1725), reflected on the suitability of different forms of administration in preparing his "Vindication of the Government of the New England Churches" (1717). Comparing monarchy, aristocracy, and democracy as possible forms of governance, Wise concludes that "the end of all good government is to cultivate and promote the happiness of all and the good for every man in all his rights, his life, liberty, estate, honor, etc., without injury or abuse done to any. . . . A company of men that shall enter into a voluntary compact, to hold all power in their own hands, thereby to use and improve their united force, wisdom, riches and strength for the common and particular good of every member, as is the nature of democracy. . . . It cannot be that this sort of government will so readily furnish those in government with an appetite or disposition to prey upon each other or embezzle the common stock" (Commager 1951, p. 122).

Here the duality of democratic existence is explicitly stated; that among the ends of government are not only the striving for the common good, but also the particular good of every member of society; the quest for conceptual continuity between the entities of social organization was initiated.

This was further elaborated in a secular context in the writings of Samuel Adams (1722–1803). A political activist, intent on extricating the colonies from oppressive exploitation by the English Parliament, he served in the Massachusetts legislature, the Continental Congress, the Massachusetts constitutional convention, and as governor. As a step in defining the place of the individual in democratic society, in 1772 he published *The Rights of the Colonists*. This document is remarkable not only because it clearly distinguishes three separate contexts of individual existence in a democracy but also because it explicitly acknowledges Locke's influence in its development. Adams asserted that the individual has a right to remain in the State of Nature with the Natural Rights to life, to liberty and to property, and to defend these rights in any way necessary and possible. The individual enters society by voluntary consent. Any Natural Right not expressly relinquished as the individual enters society remains standing. Included are the assurance of religious tolerance, a guarantee for liberty of conscience, and directions for the establishment of legislative power. To avoid arbitrary and oppressive legislation, a separate body of judges is inaugurated to assure that laws do not infringe on the individual's Natural Rights, and that the laws are applied to all individuals alike, irrespective of their state in society. Adams's *Rights of the Colonists* not only defines the liberties and responsibilities of the individual but takes a first step toward controlling the legislature by a judicial branch. The individual assumes his dual role, his natural rights, and the preservation of society as a whole.

The century-long grappling with the idea of human rights, the natural rights of the individual, found its ringing consummation in the American Declaration of Independence. There it was proclaimed "self-evident that all men are created equal, that they are endowed by their Creator with unalienable rights, that among these are Life, Liberty and the Pursuit of Happiness. That to secure these rights, Governments are instituted among Men, deriving their just powers from the consent of the governed."

Commager (1951, p. 110) hears at this point the voice of Locke: Is the Declaration in this sense an extension of the Enlightenment? Is it self-evident that the rights of the individual are unalienable, and does the Declaration give those rights the simplicity of primary qualities? Was the individual of the Declaration, defined as it was by "Life, Liberty, and the Pursuit of Happiness" comparable to the ideal systems of eighteenth century physics? Was full independence the simplifying resolution of decades of unsuccessful attempts to come to terms with the English Parliament?

Actually, the Declaration consisted of unspecific generalizations, to be implemented in accordance with a scheme of specific precepts. The Declaration transformed the colonies into sovereign states and confirmed the individual person as the ultimate unit of society. As a consequence, the colonies had to decide whether they wanted full independence from one another or instead preferred some form of association that would, at the same time, maintain each colony's uniqueness. The rights of the sociopolitical unit, whether the unit was a colony or a person, had to be balanced with the responsibilities of that unit within a larger social structure. The individual person had to be assured of his or her liberty and, at the same time, be prevented from usurping power, thus engendering factionalism. The simplicity of the Declaration had to be balanced by the complexity of the U.S. Constitution.

The journey from the Declaration to the ratified Constitution was not a smooth venture. The road was littered with printed pages of opposing views and those seeking a resting place for contemplation were jarred away by shrill arguments and counter-arguments. Just at the time of the Declaration voices had been heard asking for a form of social association corresponding in simplicity to the Declaration's definition of the individual. Thus, in his pamphlet *Common Sense* (1776), Thomas Paine denounced the complexity of the English government, and in particular attacked the parliamentary form of representation, which to many had been providing a model for the governance of the colonies. Paine declared:

> I draw my idea of the form of government from the principle in nature which no art can overturn, viz., that the more simple anything is, the less liable it is to be disordered and the easier repaired when disordered and with this maxim in view I offer a few remarks on the so much boasted constitution of England.
>
> Absolute governments have this advantage with them: that they are simple; if the people suffer, they know the head from which their suffering springs, know likewise the remedy, and are not bewildered by a variety of causes and cures. But the constitution of England is so exceedingly complex that the nation may suffer for years together without being able to discover in which part the fault lies; some will say in one and some in another, and every political physician will advise a different medicine.
>
> To say that the constitution of England is a union of three powers, reciprocally checking each other, is farcical; either the words have no meaning or they are flat contradictions (Dolbeare 1981, p. 45).

Paine's assumption that representation by a single legislative body would be adequate in the new colonies was reminiscent of the utopianism of the Enlightenment. It implied that "common sense" would avoid complexity and seek simple solutions for political reality. That the founders of the Constitution could come to an agreement about a form of government almost the opposite of a common-sense simplification is an achievement that should be fully recognized.

Actually, following the Declaration, the states entered into a loose Confederation. As specified by its Articles, a unicameral legislature was established to carry out the functions of governance. This was a response to the quest for administrative simplicity. It became apparent, however, during these crucial years, that the Confederation was inadequate for the maintenance of an army, for representation abroad, and for the proper management of finance, particularly in view of the looming financial crisis in England in 1784.

Ideas for a more effective form of governance for the associated states were proposed by John Adams (who later became the second president of the United States from 1796–1800). According to Adams, a government is instituted to provide comfort and security, the equivalents of happiness, to the greatest number of citizens. (*Thoughts on Government*, 1776, Dolbeare 1981, p. 64). Happiness also implies a concern with the principles of right and wrong conduct as the expression of morality and virtue, based on a well thought-out set of laws. In a country with a large population, the development and administration of such laws had to be accomplished by a body of elected officials, representing their respective constituencies as closely as possible. In Adams's words: "It (the body of representatives) should be in miniature an exact portrait of the people at large. It should think, feel, reason and act like them."

A single assembly was held inadequate for representing a large population, however. According to Adams, such an assembly is subject to instability, excesses of various kinds, and to corruption. As control, a second small assembly, designated as a council, is to be elected by ballot from the representatives of the initial, large assembly. This council, a precursor of the Senate in the final Constitution, and as the second house in a bicameral arrangement, was expected to exert some control of the full assembly, and to mediate between the large assembly and a separate executive branch implementing the laws and conducting negotiations (in secrecy if necessary) with foreign powers. The proprieties of legislation and of executive activity were to be scrutinized by a third branch of the government, the judiciary. Ostensibly, the administration suggested by Adams was a prototype of government by separation of powers and by checks and balances, later fully realized by the U.S. Constitution in its final form.

IMPLEMENTING CONCEPTUAL CONTINUITY IN SOCIOPOLITICAL COMPLEXITY: THE AMERICAN CONSTITUTION AND ITS INTERNAL OPPOSITES

State-Nation Oppositeness

Starting from John Adams's basic outline of a government with separate bicameral-legislative, executive, and judicial branches, fifty-five delegates representing the thirteen colonies in Philadelphia convened in February 1778, and in total secrecy during the following six months prepared a document that did not just revise, but that completely superseded the Articles of the Confederation.[4] It became the U.S. Constitution, and created a federal, national government, representing a substantial union between the states. It specified the functions of each of the three branches of the government, including administrative mechanisms for mutual controls, and implemented the principles of division of power and of checks and balances: presidential veto of proposed legislation, and overriding of veto by legislature; disapproval by the senate of presidential appointees; evaluation by a judiciary of constitutionality of laws passed by the legislature. Some of the functions of the national government were unambiguously identified—for example, maintaining an army, governing foreign affairs, controlling interstate commerce, designating tariffs, and assigning taxation in support of the general welfare. Other unspecified functions remained the responsibilities of the individual states, for example, overseeing education and militias. The detail that had gone into the Constitution became apparent where responsibilities between state and national governments were shared, as in case of introduction of Amendments into the Constitution, as pointed out by Madison in his *Federalist Paper No. 39*.[5] The reevaluation of the Constitution in the course of its ratification pointed to certain inadequacies which were then eliminated by the first ten Amendments, the first guaranteeing the freedom and rights of the individual within the National Union.

The Constitution can be perceived as a set of known elements, the three branches, and their interactions as defined variables, suggesting a "Newtonian scheme of government, static rather than dynamic."[6] Alternatively, the constitutional system may seem sufficiently undefined to provide for the flexibility and the dynamic change emphasized by Keane (1989) in the quotation included in the beginning of this chapter.

For example, that the constitutional system deliberately remains open to the introduction of variables along with its corresponding adjustments is indicated in a recent discussion of *Political Parties and American Constitutionalism* by Harvey Mansfield, Jr. (1993, p. 2):

> Besides having been formally adopted when it was ratified in 1787–1788, the American Constitution confines itself to formal statements of the powers and terms of its offices, not prescribing how they are to be exercised. Very general purposes are stated in the preamble, but they do not give specific guidance and hardly distinguish one free government from another. Thus, the Constitution leaves open the direction that policies ought to take; it does not tell the American people how it ought to live (or the American people do not tell itself how to live in the Constitution); it is more form than content. Such a Constitution of formal powers is appropriate for a government based on the "rights of man" in the Declaration of Independence, for the rights of the people, like the powers of government, are not specified as to their exercise.

Inherent in the Constitution is a set of basic opposites[7]: of the states and the nation in the public sector; of the individual and associations and corporations in the private sector; and, of the private sector (including individuals, associations, and corporations) and the public sector of the states and the nation.

Accordingly, three forms of polarization can be anticipated between these opposites. In an extreme case, opposites based on absolute principles develop as mutually exclusive, uncompromising approaches in decision-making. The struggle between the Girondists and the Jacobins during the French Revolution can be interpreted as an example of this form of opposites. During the establishment of the American Constitution the fear of the development of just such an uncompromising, factious spirit was the main reservation in the formalization of parties (see Rositter 1961; Madison *Federalist Paper # 10*). A second form of opposites represents distinctly polarized, but not irreconcilable, positions that permit a reversal of the majority-minority relationship depending on sociopolitical circumstances. A third form is the absence of a clearly defined conceptual polarization, where personalities rather than issues represent the opposite. The requirement for reconciling these opposing trends in establishing conceptual continuity became the main source of the high level of complexity characterizing the democratic form of governance. Democratic complexity is enhanced by many auxiliary deliberative units involved in government policy-making and in supervising administrative activities. Presently, the U.S. executive branch includes the presidential Cabinet, consisting of the heads of the administrative departments, including the National Security Council, the National Economic Council, and the Domestic Policy Council; a variety of agencies such as the Federal Bureau of Investigation, the Food and Drug Administration, the Environmental Protection Agency, the Health and Welfare systems and the National Endowments for the Humanities and the Arts. In the legislative process, a proposed law has to pass through deliberations by one or several committees before it reaches the floor of the House of Representatives or the Senate for further discussion and final acceptance or rejection. A veritable army of lobbyists, representing corporate and private special interest groups, tries to influence the legislative process. All this requires the maintenance of a bureaucracy of more than one hundred thousand individuals. All phases of governance are scrutinized by members of the investigative news media, who are trying to attract readers or viewers by amplifying inadequacies of subsidiary importance into sensational headlines (itself another topic entirely).

The substance of the main opposite elements was not formally given in the Constitution itself. As analogous to, but distinct from, the English parliamentary system, opposites were represented by the main political parties that became characteristic components of the democratic process as it evolved in the United States (Commager 1951, p. 192; Ladd 1970, 1975; Kessler 1993, p. 230; Schramm and Wilson 1993). At different times, the parties indicated the two sides of the basic opposites, but at other times the apparent sharpness of their outlines was blurred. In this role, the parties largely contributed to the dynamics of the democratic process. As a conspicuous (but not necessarily desirable) characteristic, the parties generated a striving for power in the American political system and imbued that system with its adversarial-competitive-majoritarian complexion.

The changeableness in the approach of the parties to these opposites at times opened a path to desirable adjustments between diverging attitudes; but at other times it was perceived as threatening uncertainty. It is this changeableness, Keane's "flexibility" or Berlin's "uneasy equilibrium" that emphasizes the temporal dimension of the Constitution giving it its full measure of complexity. This is substantiated by a brief account of the continuities and changes in the party system.

The party system originated during the initial period of nationhood as a factor in the development of the state-nation oppositeness. Promotion of the National Union under the presidencies of George Washington (1789–1797) and John Adams (1797–1801) became the main aim of the Federalist party (later transformed into the Whig party). Strengthening the rights of the of states and the promotion of decentralization under the presidencies of Jefferson (1801–1809), James Madison (1809–1817), and James Monroe (1817–1825) became the main cause of the Republican-Democratic party. Under the guidance of Alexander Hamilton, the initial concern of the Federalist party was to revamp the financial status of the nation by introducing tariffs and a limited taxation on whiskey. These actions were aimed at eliminating the national debt, incurred during the revolutionary war. In spite of much controversy, a National Bank was created to support the development of a business community. The bank later lost popularity because its conservative policies interfered with the risky lending policies of state-chartered banks. The Federalist-Whig party was supported by members of the commercial class, which included maritime industries, shipowners, fishermen, whalers, wealthy planters, and part of the common citizenry. By the introduction of the Alien and Sedition Acts, under John Adams's presidency, there were attempts to prevent disruptive anti-national and pro-state propaganda, incited by the French Revolution. As a potential threat to the new nation's freedom of speech, the enforcement of the Alien and Sedition Acts became a major issue in the contest between Federalist and Republican-Democratic parties, the latter, having condemned the Acts, gained control of the government with Thomas Jefferson as its president. Expansions of trade and manufacture on the one hand, and the growth of agriculture (including cotton growing in the South) on the other began to widen the gulf between Federalist tendencies toward centralization and the Republican-Democratic goals of decentralization. The rights of the states were reasserted in the Kentucky and Virginia resolutions.

Inconsistencies in the development and complexities of the state-nation oppositeness were already appearing during the first periods of Republican-Democratic party rule. Even though he was a chief protagonist of states rights and opposed to expansion of the national government, Jefferson had to use national power to bring about the Louisiana Purchase and to control shipping during the Napoleonic wars with England. Similarly, mobilization of the army for the war against England became a national endeavor under Jefferson's states rights successor, James Madison. As well, under Madison's presidency, Chief Justice John Marshall (who had been appointed by the Federalist President John Adams) interpreted the Constitution as favoring national rather than state interests in several Supreme Court decisions.

Following the war with England, the Republican-Democratic party retained power but following a Report on Manufacture (written by the Federalist Alexander Hamilton) Federalist programs became the basis of the national economy. The National Bank, discontinued earlier by Jefferson in 1811, had to be rechartered in 1816 by the Republican-Democratic party under President James

Madison. A program for the construction of key roads and canals was also inaugurated as a national venture in spite of the states' rights orientation of the Republican-Democratic party. The introduction of the Monroe Doctrine was a further expression of national interests that closed the American continent to further European colonization (for review see Capers 1989).

An economic crisis beginning during the Monroe presidency (1817–1825) led to disarray in the party system. The Federalist party continued, now as the Whig party, and the Republican-Democratic party became simply the Democratic party. The slavery problem especially, and sectionalism between East, West, and South assumed prominence. The slavery issue led to the Missouri Compromise of 1820. The initial proposal, prohibiting the introduction of slavery into Missouri and emancipation of all children born of slave parents at age fifteen, was not passed by the Senate. A temporary resolution was achieved by admitting Maine into the Union as a free state and Missouri as a slave state; but slavery was prohibited in the Louisiana Territory (Capers 1989). Disparity of sectional interests became even more evident when Andrew Jackson became president (1829–1837). Jackson had to reconcile subsidies for development of the West with the Northeast's insistence on reducing Federal expenditures; favored high tariffs by the industrializing Northeast and the need for lower tariffs in the South; and the states' attempts to establish their right to nullify Federal laws (tariffs in particular), against the clamor of the West for stronger action by the national government. Rather than facing the complexity of these issues, Jackson directed his attention to the contentious issue of the National Bank, and vetoed its recharter as a simplifying solution. Jackson's "solution" set an example for escaping from complexity by substituting the simplicity of a resolvable problem.

Under the leadership of John C. Calhoun, the South's initial demand for expansion of states rights initiated a polarization, with Jackson rejecting this demand: "Declaring the federal government sovereign and indivisible, denying that any state could refuse to obey the law, and rejecting the notion that any state could leave the Union" (quote from R. M. Goldman 1989). But Congress empowered Jackson to enforce the law prohibiting secession by military action. This possibility of a simplistic military solution was temporarily averted by the Van Buren administration immediately following (1837–1841). State jurisdiction over slavery was admitted within each State's territory. However, tensions increased with the westward expansion, raising the question of slavery in the newly added states.

In the interval from the end of Jackson's presidency in 1836 to the election of

Lincoln in 1860, increased divergence between North and South was prevented by camouflaging the divisive problems in ambiguities and by repeated concessions to both abolitionists and secessionists (Capers 1989). For some time Lincoln himself was not an unconditional abolitionist: to him the continuation of slavery in certain states, and the prospect of its gradual elimination, seemed compatible with the maintenance of the Union. But demands by the South for elimination of all restrictions on slavery in the newly added territories and the secession of six of the southern states were unacceptable threats to the Union. Under these conditions, Lincoln could not yield to the South's demands for withdrawal of northern army units from Fort Sumter in the Charleston Harbor, provoking the shelling of the fort by the Confederate Army, the first military action of the Civil War.

The bombardment of Fort Sumter represented the abandonment of the continued search for compromise in light of the development of economic interdependence between southern cotton and the northern textile mills, and the eventual abandonment of slavery. Compromise, the possibility of conceptual continuity, was replaced by military action, the last and lowest vestige of the inability of the primitive human mind to cope with complexity. It was a "solution" to a state-nation dichotomy that cost more than 600,000 lives and caused extensive destruction. As with most military actions, such "solutions" are simplifications of complexity that carry a high price tag in human suffering. For the African American people, the end of the Civil War only opened the struggle for civil rights that, after more than a hundred years, is still incomplete. Achieving equality and elimination of racism and the adjustments in the state-federal relationship became distinctive requirements of sociopolitical conceptual continuity.

The Private/Public Oppositeness

With the end of the Civil War, the prominence of the state-nation oppositeness declined, and concern with the individual as both "private egoist and public-spirited citizen" (Keane 1989), the impact of private interest groups on the public system, and restraints of the individual by the government became the main issues in American sociopolitics. The vagaries of private/public oppositeness are due to its incomplete definition in the Constitution, as referred to earlier. As understood here, private/public oppositeness demonstrates the source and nature of the complexity inherent in democracy even more clearly than state/nation oppositeness. In speaking about vagaries of oppositeness, it is meant that at different times one or the other of the opposites predominates or their adversarial positions interfere with effective governance. This is regarded as the polarized state of the oppositeness. Alternatively, a depolarization can take place by an integration of at least some aspects of the opposites. Shifting from one form of the oppositeness to another, polarized —depolarized, introduced a dynamic momentum into the democratic system that further increased its complexity. Management of this dynamic phase became the main function of the political parties (for review see Davis 1989).

Here again, the private-public oppositeness emerged in Jefferson's time when the newly united colonies seemed to offer to the private individual—as farmer, as craftsman, or as professional—unlimited opportunities to achieve an adequate livelihood without subsidy from the public sector. These opportunities for personal rewards had to be protected from repressive intrusion by the government, and hence the private-public oppositeness was shifted in favor of its private component. The hands-off policy on government intervention was upheld initially even when the results of private initiative were detrimental to the economy. Thus, uncontrolled land speculation during the first part of the nineteenth century led to a severe economic depression and a call for government adjustments on behalf of those most severely affected. At that time, however, President Van Buren rejected the idea of the government's responsibility to rectify the supposedly transient failure of the economics of the private sector (Van Buren, Message to Congress, Commager 1951, p. 323). Ample opportunities for starting new enterprises in the expanding nation prevented such economic crises from leaving lasting scars. The opening frontiers invited the bold and often reckless economic daring that generated the transcontinental railroads, steel-mills, mining, and later on, even agriculture on a large scale. The eventual achievement of an unprecedented level of productivity was hailed as the result of individual enterprise, of laissez-faire economy.[8]

The shifting of the private-public oppositeness to the private side was accentuated during the second half of the nineteenth century and into the first decade of the twentieth. It was championed by the Republican Party and those of its presidents who held political power in that period, except under the terms of Grover Cleveland's Democratic administration (1885–1889; 1893–1897).

During the latter part of this period laissez-faire competition was reinforced by the popularization of Herbert Spencer's sociological evolutionary theories, with their emphasis on the survival of the fittest. Spencerian thought became embodied in the ideology of William Graham Sumner (1840–1910), who, as chairman of the newly created department of American Political Science and Sociology at Yale, became one of the founders of these disciplines. An excerpt from his *The Challenge of Facts* (1880) indicates his attitude toward the private-public oppositeness (from Commager, 1951, pp. 328–329):

> The condition for the complete and regular action of the force of competition is liberty. Liberty means the security given to each man that, if he employs his energies to sustain the struggle on behalf of himself and those he cares for, he shall dispose of the product exclusively as he chooses.
>
> What we mean by liberty is civil liberty, or liberty under law; and this means the guarantees of law that a man shall not be interfered with while using his own powers for his own welfare.
>
> Liberty, therefore, does not by any means do away with the struggle for existence. We might as well try to do away with the need for eating, for that would, in effect, be the same thing. What civil liberty does is to turn the competition of man with man from violence and brute force into an industrial competition under which men vie with one another for the acquisition of material goods by industry, energy, skills, frugality, prudence, temperance, and other industrial virtues. Under this changed order of things the inequalities are not done away with . . . but it is now the man of the highest training and not the man of the heaviest fist who gains the highest reward. It follows from what we have observed that it is the utmost folly to denounce capital. To do so is to undermine civilization, for capital is the first requisite of every social gain, educational, ecclesiastical, political, aesthetic, or other.

Laissez-faire economic polity was protected in the courts, and laws that apparently infringed on the freedom of individuals to act according to their advantage were voided. For example, a New York State law prohibiting the making of cigars in tenement houses was declared unconstitutional in 1885 by a higher court. Referring to this law, Justice Earl declared that "such governmental interferences disturb the normal adjustments of the social fabric, and usually derange the delicate and complicated machinery of industry and cause a score of ills while attempting the removal of one" (Commager 1951, p. 332).

Two additional instances substantiated the broad protection of the laissez-faire principle by major Supreme Court decisions. An initial attempt to introduce antitrust legislation (the Sherman Antitrust Act, 1890) was weakened by the Supreme Court, who restricted it to trade, but exempted manufacture. The Supreme Court also rejected the introduction of an income tax proposed by President Cleveland in 1894.

A more controversial congressional act, one setting minimum wages for women, was dismissed at first hearing. The intricacies of the decision are indicated by the two sections of Justice Sutherland's opinion in this case (1923). It is one of those Court rulings that indicates the need to consider a multiplicity of factors in the social system and to evaluate the significance of those factors in a particular court case. Further, it emphasizes that the laws maintaining a civil society only have restricted general validity, and that it is necessary to consider each case within its specific setting and circumstances. Justice Sutherland's dismissal (1923) was reversed as mentioned later on. The ruling is extensively quoted here, because a comparison with its rejection at a later date (see below) demonstrates the delicate shifts in Supreme Court rulings pertaining to the private and public sectors.

> There is, of course, no such thing as absolute freedom of contract (e.g., in respect to minimal wages). It is subject to a great variety of restraints. But freedom of contract is, nevertheless, the general rule and restraint the exception; and the exercise of legislative authority to abridge it can be justified only by the existence of exceptional circumstances. Whether these circumstances exist in the present state is the question to be answered.
>
> The liberty of the individual to do as he pleases, even in innocent matters, is not absolute. It must frequently yield to the common good, and the line beyond which the power of interference may not be pressed is neither definite nor unalterable but may be made to move, within limits not well defined, with changing need and circumstance. Any attempt to fix rigid boundaries would be unwise and futile. But, nevertheless, there are limits to the power, and when these have been passed, it becomes the plain duty of the courts in the proper exercise of their authority to so declare. To sustain the individual freedom of action contemplated by the Constitution, is not to strike down the common good but to exalt it; for surely the good of society as a whole cannot be better served than by the preservation against arbitrary restraint of the liberties of its constituent members (Commager 1951, p. 333–334).

After a transient shift toward the public sector under the presidencies of Theodore Roosevelt and Woodrow Wilson between 1904–1920, the laissez-faire promotion of the private sector reasserted itself. Under the successive four-year terms of Presidents Harding, Coolidge, and Hoover (1920–1932), the private sector, dominated by big business and supported both by Republican legislatures and presidencies, once more assumed a leading role. Congress enacted a high tariff (Fordney-McCumber Tariff, 1922) that protected not only the major industries but agriculture as well. Taxes were lowered and consolidation of businesses was promoted. Meanwhile, the judicial branch avoided enforcing the existing antitrust laws. An unparalleled expansion of industry included the production of automobiles, chemicals, electric machinery, and appliances, together with the development of electric power, highway construction, a building boom, and the development and spread of advertising and commercial broadcasting.

Prohibition and anti-prohibition, social and cultural intolerance, exclusion of ethnic and racial minorities from the social mainstream, rejection of the teaching of evolution, a flourishing Ku Klux Klan, an advance in the emancipation of women with the granting of voting rights, and a heightened pursuit of superficial pleasures were the characteristics of the 1920s. The period ended with the stock market crash in 1929 and the subsequent catastrophic depression that, for several decades, curtailed the role of laissez-faire and the concomitant politics as the undisputed characteristic of American nationhood.

Even before that juncture, private sector individualism and the laissez-faire economy did not remain unchecked. The rapid growth of industrial powers during the second half of the nineteenth century raised the question of whether the resulting increase in profits should remain unchallenged. When U.S. railroads began to charge excessive rates for grain storage in the railroad-owned warehouses, several states imposed limits on those charges. The railroads promptly challenged this legislation as an instance of "deprivation of property without due process of law" (Commager 1951, p. 325). The challenge was dismissed on the ground that a private property that affects the broad public interest, such as basic staples like bread and other grain products, is in a legal sense no longer a purely private property, and therefore the price of storage was to become regulated in the public interest (*Munn versus Illinois*, 1876. Commager 1951, p. 327).

Later on, when laissez-faire theory was invoked to prevent restrictions on working hours in bakeries, the restrictions were upheld by a Supreme Court majority. In his dissenting opinion, Justice Oliver Wendell Holmes initiated anti-laissez-faire jurisdiction (*Lochner versus State of New York*, 1905. Commager 1951, p. 342).[9] Later on, by endorsing Justice Holmes's dissenting opinion the laissez-faire stand against wage regulations for women was overturned on appeal, confirming the decision of the Supreme Court of the State of Washington to uphold the right of the state to control minimum wage practices. In this instance, the interpretation of liberty by Justice C. J. Hughes, in favor of the public sector, is of particular relevance because it is in opposition to the opinion of Judge Sutherland quoted earlier. A comparison of the actual wording of the two opinions indicates the precariousness of the balance favoring either the private or the public sector in accordance with the U.S. Constitution. Justice Hughes stated:

In each case the violation alleged by those attacking minimum wage regulation for women is deprivation of freedom of contract. What is freedom? The Constitution does not speak of freedom of contract. It speaks of liberty and prohibits the deprivation of liberty without due process of law. In prohibiting this deprivation the Constitution does *not recognize an absolute and uncontrollable liberty*. Liberty in each of its phases has its history and its connotation. But the liberty safe-guarded is *liberty in a social organization* which requires the protection of law against the evils which menace the health, safety, morals and welfare of the people. Liberty under the Constitution is thus necessarily subject to the restraints of due process, and regulation which is reasonable in relation to its subject and is adopted in the interests of the community.

What can be closer to the public interest than the health of women and their protection from unscrupulous and overreaching employers? And if the protection of women is a legitimate end of the exercise of state power, how can it be said that the requirement of the payment of a minimum wage fairly fixed in order to meet the very necessities of existence is not an admissible means to that end? (Washington Minimum Wage Case, 1937. Commager 1951, p. 334).

The continuity of laissez-faire had already begun to falter toward the very end of the preceding century; high profits, coupled with the misery of the labor force, created a social disparity that led to populist and socialist movements and called for some form of readjustment. Thus, as a counterbalance to laissez-faire ideology and as a basis for extending governmental regulatory activities, a "Theory of Public Interest" appeared (Davis 1989) under the progressive Republican presidency of Theodore Roosevelt. In this context, the Interstate Commerce Commission was created (1887) as a regulatory agency, and the Pure Food and Drug Act and Meat Inspection Act were passed and signed into law (1906). The public domain was increased with the establishment of national forests, and related reclamation and conservation projects. After a short hiatus during the Taft administration (1908–1913) promotion of the public sector was vigorously continued under the Democratic presidency of Woodrow Wilson (1913–1921) who won the elections against Republican tickets divided between conservative and more progressive forces. The protectionist tariff was reduced, an income tax to raise revenues was instituted, and a system of Federal Reserve Banks was established as protection against economic instability of the private sector. A new Federal Trade Commission issued orders against unfair business practices and enforced antitrust proceedings. Legislation was also introduced to promote the interests of labor. Toward the end of Wilson's presidency, participation in the First World War became a unifying cause for the country, and matters of international politics became the dominant issue.

The period after the First World War and the Wilson era brought a temporary resurgence of laissez-faire policy. The economic disarray at the end of the laissez-faire postlude of 1920–1932 under the Republican presidencies of Harding, Coolidge, and Hoover led to the takeover of political power by the Democratic party with Franklin Delano Roosevelt's introduction of the "New Deal" as a political philosophy, under a radically divergent interpretation of the Constitution (for review see Roohan, 1989).[10] Man's unchangeable natural rights were replaced by rights that were products of sociopolitical evolution. The flexible conception of human rights, adaptable to changing social conditions, was emphasized. This was a reinterpretation that attempted to "transform a regime based on natural rights into one that could secure real, evolving, social rights. The new theory taught that rights emerged from the social contract, that they were defined by a bargain struck between the government and the governed, which bargain was subject to continual renegotiation as circumstances altered" (Kesler 1993, p. 235). In this view, people struggling to produce the bare necessities to support their existence cannot be free. Therefore, it becomes the government's responsibility to provide opportunities for training and education as a first step toward one's entry into gainful work, and to relieve one's fears of nonaffordable medical care and of destitution in old age (p. 236). These newly assumed responsibilities were met by the introduction through the New Deal of "social entitlements" with a corresponding expansion of the central bureaucratic administration.[11]

The New Deal policies favoring the public sector prevailed without decisive opposition until 1968 (for review see Hamby 1989). Even President Eisenhower, as a Republican, had to compromise with the New Deal program of a Democratic Congress by endorsing major public works such as the Interstate Highway System. Eisenhower's presidency is noteworthy by achieving sufficient consensus between the parties to maintain administrative efficiency. In contrast, the later Republican presidents (Nixon, Ford, Reagan, Bush) reactivated private sector policies against the background of the public sector oriented Democratic Congress. This generated an ideological, polarizing shift in the private-public duality with the reduction of the regulatory powers of government. It lead to the deterioration of the savings and loan banking system and large-scale financial transactions (junk bond market) which did not generate investments and savings for the maintenance of global industrial competitiveness. Expanding government expenditures without corresponding growth in revenues increased the national budget deficit and the national debt.[12] In a reverse situation, polarization of the budget debate between Clinton, as Democratic president and an uncompromising congress, dominated by a Republican majority, ideologically committed to the fostering of the private sector interests, led to two shut-downs of governmental functions.

The citizen's increasing dissatisfaction with the inadequate conduct of governance became manifest in a decline of voting participation and the surprising support of Ross Perot as an independent presidential candidate and of other radical simplifiers as potential political and religious leaders.[13]

The growth and development of an American democracy into dynamic polarizations of sociopolitical reality is regarded here as an attempt to see society not in its full complexity, but to cope alternately with only one of its components. The more extreme the polarization, the simpler the represented reality, but the more difficult it becomes to reconcile the opposites and to establish conceptual continuity between the factions of the social system as a whole. This polarization is enhanced by the elections in which the issues are presented in the forms of simple ideas and terms that are easily grasped by the voting public. After the elections, the public expects the promised simplifications to be implemented, even though the problems to be dealt with require complex solutions. Projecting a simple reality to the public while actually trying to cope with a highly complex reality is one of the difficult tasks of the American Presidency.

The recurrence of the political polarization in the course of American governance is paralleled by the liberal and communitarian representation of the individual in the academic thinking of the twentieth century (Lukes 1973a,b). In the liberal view the individual is seen as a legalistically isolated entity that creates secondarily the values of society (ontological individualism, procedural individualism; Rawls 1971; Carens 1993a,b). The individual is endowed with innate rights with a neutrality that avoids intrusion of moral values. In this form, the individual is atomistic in the sense of irreducible "oneness" that neglects those social relationships that are essential characteristics of individual actuality. Individuals are "separate entities rigidly bounded and impermeable to each other" (Rosenblum 1993, p. 78).[14] The laissez-faire individual belongs to this category of liberal individualism. Schumpeter (1987) has advocated constructing an economic theory on the basis of self-interest alone. Riesman's (1961) individuals attempt to overcome their isolation by joining the "lonely crowd."

In the opposite view, the common good takes precedence over individual independence (holism, communitarianism, civic republicanism). Here, relatedness between individuals, engendered by shared social experiences and expectations, and between individuals and society becomes the main characteristic of the individual. The shift toward the communitarian side of individualism in the American scene was reinforced by a revival of John Dewey's system of thought (1927, 1929; Westbrook 1992; Menand 1992), by Sandel (1996), by Bellah et al. (1985, 1990). Also, Habermas's writings greatly stimulated communitarian thinking in the Western world in general (White 1995). In a separate trend the diversity of social groups, the relationship between them and between the individual and the diverse groups, are being explored in the studies of pluralism (Kekes 1993; Eisenberg 1995).

In an additional development, the individual restrains self-interest for moral reasons. The American sources of morality are the Ten Commandments of the Old Testament; the Christian call for charity, and its objections to excesses of self-serving individualism; and the faith in the perseverance and progress of humanity and individual achievement, manifest as private property—all under a limited government. The American form of moral individualism has deep roots in the tradition of religious congregationalism. In this particular form, "moral" individualism is thought to play a significant role in the stabilization of American democracy.[15] This position is supported by an exceptionally high level of religiosity of the U.S. population (about 70 percent) as compared with the much lower levels (30–50 percent) in European countries with equally developed economies (Ladd 1993b; 1993–1994).

A particular reason for including the moral individual is the proposition (Gutmann and Thompson 1996) that in a democracy moral conflicts cannot be avoided. Some moral opposites (abortion, pro-life) preclude compromise, and hence prevent the establishment of conceptual continuity. Such moral positions can be thought to be as irreconcilable as ideologies.

At the end of the twentieth century, any evaluation of American democracy may have to consider several contrary phases in its existence. On the positive side, American democracy survived the ordeal of a devastating civil war, and participated decisively in two world wars, using thermonuclear bombs for the first and so far only time in ending the second. Through its Marshall Plan, the United States also assumed nearly full responsibility for the restoration of its European adversaries; it introduced democratic governance in Japan; and it eventually prevailed in the Cold War against Soviet communism.

On the negative side, the United States failed diplomatically and militarily in its confrontation with Vietnam, with the corresponding erosion of its morale. The efficiency of the domestic majoritarian-competitive system became impaired by adversarial confrontations, extensive periods of governmental gridlock and dishonesty (e.g., Watergate, the Iran-Contra Affair), the Whitewater Scandals, the 1996 election campaign irregularities, and by the ensuing decline in voter turn-out.[16]

At the beginning of the second half of this century, the United States was the main global economic creditor, but it is ending the century as the main global debtor. It made significant progress toward implementing civil rights, but the inequality of living standards between the strata of its population increased dramatically. The initially high competitiveness of its industries declined. Regaining competitiveness required increased productivity and managerial reorganization that has resulted in extensive lay-offs and the prospect of permanent unemployment for certain segments of the labor force. However, these negative phases proved to be transient, and subsequent adjustments were followed by a remarkable strengthening of the economy.

At this point, three alternatives can be envisioned for the future of democracy. The extension of this chapter points to the consensual, non-competitive executive component of the Swiss democracy. As a major threat to democracy, the following chapter gives a brief account of two forms of totalitarianism (Chapter 6). Deliberative democracy is then invoked in Chapter 7 as a strong antidote to democratic decay and totalitarian aggression, and as implementation of conceptual continuity.

THE SWISS MODEL OF CONSENSUAL DEMOCRATIC COMPLEXITY

Switzerland often claims to be the oldest democracy, and dates its formation from 1291 with the alliance of three communal districts later known as the cantons of Schwyz, Uri, and Unterwalden. That period in Swiss history actually had little continuity with the more recent unification of the population of modern Switzerland by the establishment of the definitive and complex Swiss Constitution of 1848. But, if 1291 is taken as the beginning of Swiss self-government it merely indicates the long and tortuous process required to reach the final goal. Indeed, just because it took Switzerland all that time to overcome a long series of sociopolitical hurdles, its history may be a role model for the establishment of democratic institutions and a source for the study of the success and failure of conflict resolution on the path to democracy. The history of Switzerland is briefly recapitulated here as a different example of the conversion of a simple sociopolitical reality into a complex one (Brooks 1927; Codding 1983; Diem 1993).

In its history, the area of contemporary Switzerland was an outpost of the

Roman Empire for some time; that occupation was followed by invasions of German tribes of the Ostrogoths and Franks. Christian missionaries also arrived at about the sixth century and set up abbeys at St. Gallen, Zurich, Disentis, and other places. The area eventually became part of the empire of Charlemagne and was later divided between his sons. In 1033, the area was integrated into the Holy Roman Empire under the rule of feudal lords.

Essentially, the 1291 alliance between the three cantons assured mutual military aid as a first step toward the liberation from exploitation and oppression by the feudal powers, particularly the Habsburgs, who dominated large sections of the area. An initial military campaign by the new alliance ended with the defeat of the Habsburgs in 1315 at Morgarten, and a temporary truce. In an advance, important for the development of the Swiss Confederation, the alliance was joined by three main cities of the area, Lucerne, Bern, and Zurich. The last supported highly developed guilds of artisans and superior means of defense. In spite of the truce, several military forays by the Habsburg forces against different parts of the alliance were beaten back; Zug and Glarus next joined the alliance; and by the end of the fourteenth century the Confederation had grown to eight cantons.

The military confrontations between the Confederation and the Habsburg armies and allies continued through the fifteenth century, with Zurich siding at one time with the Habsburgs. Military actions were brought to a temporary end in 1446 by the Peace of Constance, and Zurich was brought back into the Confederation. Meanwhile, collateral aggressive moves by Burgundy ended with the defeat of the Burgundian army at Morat in 1476. A last attempt to reestablish Habsburg domination of the Swiss by Emperor Maximilian I of Austria (mainly as source of tax revenue) resulted in the defeat of the Habsburg armies and a peace settlement in 1499, which, after 200 years of strife, finally put an end to Habsburg encroachment. The Swiss Confederation now reached its peak of military power, capped, however, by the defeat of its military expansion into northern Italy.

Nevertheless, with the defeat of the Habsburgs and Burgundians the status of Switzerland as a military power became sufficiently prominent to avert new external military threats to its safety at least temporarily. However, the removal of external dangers abolished the main aim and purpose of the Swiss alliance and internal discord increased. There were rivalries between the major cities, in particular Zurich and Bern, and another fundamental animosity between urban and rural districts. Severe antagonisms erupted with the progress of the Protestant Reformation, and Zwingli and Calvin assumed strong leaderships of

Swiss protestantism. These internecine squabbles and confrontations need not be identified in detail, but it should be noted that the threat of a common enemy was the simplifying condition promoting the unification of some of the diverse elements in the region during the first two hundred years (1291–1499) of Swiss history. In contrast, and in the absence of this overriding military threat during the next 200 years, disunity and dissension became dominant, and the increase in the number of cantons in the Alliance enhanced that trend. Cooperation between the cantons remained only a matter of military expediency; each of the cantons maintained its strict sovereignty and the complete loyalty of its inhabitants. In this context, to regard Switzerland as a democratic confederation during this early period of its history, and to speak of it as the oldest democracy, may seem objectionable.

The crucial transformation of Switzerland from a military confederation into a nation of cantons with common civilian interests occurred between the end of the eighteenth and the middle of the nineteenth centuries. Even the final steps in the creation of an advanced democracy required more time than the corresponding period in American history. This may be relevant in dealing with similar transformations, for example, the current situation in Russia or in some of the newly emerging nations of the Third World.

As late as 1775, Swiss patriotism in terms of willingness to sacrifice property and life was still centered on the individual cantons and not on Switzerland as a nation (Kohn 1956, p. 15). The need to unify the cantons was promulgated then by members of the University of Basel, who in 1760 had established a Helvetic Society with the aim of promoting the spirit of Swiss national unity. This effort toward unification initially found only a literary expression without practical political consequences, and was probably a step toward nationalistic simplification rather than toward democratic complexity. But the Helvetic Society did have an educational effect, and prepared some of its members for the ensuing sociopolitical changes.

Despite the military alliances between its cantons, Switzerland was overrun by the French armies during the Napoleonic conquest of Europe (1789). Under the auspices of the French Directorate the Swiss cantons were unified for the first time with the inauguration of the Helvetic Republic (1798). A legislative body elected the five members of the executive unit. The executives delegated administrators to the cantons to oversee the national interests in creating a centralized government, and a series of political changes were introduced to remove some of the traditional Swiss impediments to democratic governance, in particular the entrenched oligarchies.

The Helvetic Constitution of 1789 abolished all special privileges and established the equality of individuals and of territories. For the first time it created a common Swiss nationality and removed all the many internal barriers to trade and intercourse (Kohn 1956, p. 40).

Following the French model, the Helvetic Constitution emphasized centralization and an authoritarian executive power; but those proved incompatible with the Swiss penchant for local government and diversity (Kohn 1956, p. 39). There followed a period of great unrest, characterized by attempts to reconcile the tendencies to unification and equalization of the citizens of all cantons with their attachment of their traditional parochialism. These sweeping reforms, together with French confiscation of large portions of the Swiss treasury prompted insurrections against the French. In some instances they were squashed by ruthless massacre of men, women, and children as in Stanz in 1798. Switzerland's situation was worsened because of its geographic position. After the outbreak of the war between France and Austria, Switzerland became the favored battleground for encounters between their two armies. The resulting impasse of loyalties was resolved by Napoleon, who ordered the Swiss people "to abandon their petty factionalism and to devote themselves to the common needs of the fatherland" (Kohn 1956, p.44). The constitutional reforms imposed by Napoleon (1806) attempted to balance the need of a modern state with liberty and equality against traditional values (see Kohn 1956, p. 46).

The modified constitution recognized nineteen cantons, each with its own administrative body. The original name "Confederation" was reintroduced, and a legislative body was established. This legislature was now composed of two representatives from each of the larger cantons and one representative each from the smaller one. Meetings of this legislature were held in turn in the six largest cantons with the chief official of each canton being the temporary President of Switzerland. There was considerable admiration for Napoleon. Even after the French leader's defeat in Russia in 1813; the *Landamann* (President) of Switzerland praised Napoleon for his mediating the establishment of a stable federal structure and public peace, "which had reconciled the most bitter animosities" (a typical case of establishment of conceptual continuity in a sociopolitical system). Also, the German-speaking Swiss did not share the chauvinism rampant in Germany after Napoleon's defeat, and exhibited a more pragmatic attitude to a new national self-identity. The attempt by Napolean to impose a foundation for the modern Swiss state via his foreign power (Grüne and Pitterle 1983, p. 32) may be regarded as a parallel of contemporary attempts by the United States to introduce democracy in underdeveloped countries.

With the decline of Napoleonic power, some of the old reactionary interest groups attempted to reassert themselves, resulting in the Pact of 1815. In it, the legislature was reduced to one representative from each of twenty-two cantons, with the chief executives of each of the three cantons—Zurich, Bern, and Luzern—becoming presidents for two years at a time.

During this period the inhabitants of French and Italian districts immediately adjoining Switzerland expressed their desire to join the Swiss Confederation. This led to the establishment of Italian- and French-language cantons and to the introduction of trilingual communication for official purposes. Also, after Napoleon's defeat in 1813 perpetual neutrality was confirmed at the Congress of Vienna and in the Treaty of Paris in 1815; but the pact of 1815 was an agreement between cantonal governments, not a decision by the Swiss people themselves, to whom it was not submitted for discussion or approval. The 1815 Constitution had not yet established a common citizenship: it did not introduce a common currency, a national postal service, a trade agreement between the cantons, or even specifications for a common railroad system. Once again, unification referred only to military defense (Kohn 1956, pp. 57–58).

In view of the deficiencies of the 1815 Pact, additional steps were required to further unity of the cantons. For about fifteen years (1815–1830), the idea of a broader unification remained not only a source of indecision but an source of increasing polarization between the prominently Catholic rural cantons, who took a conservative stand for canton sovereignty, and the urban Protestant centers, who were striving for a radical revision of the existing pact and the inauguration of a truly democratic federal system. Agitation led by refugees from uprisings in Italy and Germany in 1830 added to the polarizing political situation in Switzerland, and resulted in the full expression of the contrasts between liberal and conservative, radical and clerical, and industrial and agrarian parts of the Swiss population (Kohn 1956, p. 76). In particular, religious strife intensified between the Protestant and Catholic cantons. The situation was aggravated by the emotional and ideological identification with the German and Italian nationalistic movements by the corresponding parts of the Swiss population. These developments culminated in a final confrontation between the predominantly Protestant industrialized and Catholic agrarian cantons, the latter of whom had established the *Sonderbund,* an ideological alliance. After an initial failure to reach any compromise agreement for cooperation, the *Sonderbund* cantons faced threats of military action. Their defeat in a short campaign, conducted as humanly as possible, by the Genevan General Dufour, created conditions for reconciliation and efforts toward constitutional reforms were joined (transition from conceptual discontinuity to conceptual continuity).

In 1848 a definitive federal constitution was adopted, and later revised in 1874. Its 123 articles defined "the rights and duties of the citizen and of the governing bodies" (Diem 1993). As a political system, Switzerland became a federation that now comprises about 3,000 communes distributed among 26 cantons (including six half-cantons); all of these cantonal units maintain a high degree of autonomy.

Although the Swiss constitution follows the American model in part, certain basic deviations give it a distinctive character. The Swiss Constitution establishes a bicameral legislative body, the *Federal Assembly,* consisting of a *Council of States* corresponding to the U.S. Senate, and a *National Council,* corresponding to the U.S. House. In the Council of States each half-canton and canton is equally represented by one or two delegates, respectively. In the National Council each elected member represents 22,000 citizens, giving those cantons with higher populations a proportionally larger number of representatives. The two chambers have equal law-making authority (Penniman 1983, p. x). Candidates for membership in either one of the two chambers are nominated by the political parties of each canton and are elected by popular vote. Accordingly, partisanship is more pronounced on the cantonal level than on the federal level.

The Federal Council, the executive body of the Swiss government, departs radically from the American model. The Council consists of seven members, each chosen for a three-year term by the Federal Assembly from among its membership. As a rule, positions on the Federal Council are filled by legislators with experience in government. Members of the Federal Council designate one of their number as council chairman and Federal President for a one-year term. Members of the Council and its President can be reelected to any number of terms. The Federal President represents the nation at home and abroad and acts as an overseer of the Council's activities. A Vice President is chosen at the same time and in the same manner as the President. Every year each member of the Federal Council is appointed to assume responsibility for the efficient functioning of one of the main government departments (Brooks 1927, p. 118).[17]

The Federal Council reflects the pluralism of Swiss society, for its members must include representatives of the four dominant political parties. As of 1983, the Council consists of two representatives, each of the Radical, Catholic, and Social Democratic Parties and one representative of the Popular party. This composition referred to by the Swiss as the magic formula has remained essentially unchanged, even though the strength of the parties has fluctuated over the years. In addition, each of the Council members must belong to a different canton, and representation of the linguistic minorities and the two major religious denominations is required, along with the cantons of Bern and Zurich.

The Council deals with many diverse administrative matters, and introduces important legislation in the Federal Assembly, where it can be confirmed or rejected. In contrast to American legislative process, the executive Council cannot veto legislation by the legislative branch Assembly. Policy decisions of the Council are transmitted to the public as expressions of single joint opinions of the whole Council. This can indicate consensus of the Council members, called a consociational decision. If consensus is not reached, the divergence is resolved by a suitable interpretation representing a common ground; these rulings are called "interpretative decisions" (Steiner 1983). Therefore, the Federal Council appears as a consensual, non-majoritarian, non-adversarial, An alternate mechanism for legislative decision-making in modern Switzerland is the popular referendum, an extension of a medieval tradition (Codding 1983, p. 18; Brooks, 1927). Its frequent use is another distinct characteristic of Swiss democracy incorporated in the 1848 Constitution. As few as seven voting citizens can request a referendum by submitting an outline of the desired change. If the request is approved, a vote will be called when 50,000 signatures are obtained in support of referenda on ordinary legislation and 100,000 signatures in support of referenda on constitutional amendments. Also, newly proposed laws must be submitted to a public referendum if so requested by 60,000 signatures from eight cantons. In five cantons, legislative authority is assumed by direct democracy in form of an open meeting, the *Landtag*.

The Swiss judicial branch takes the form of a Federal Court that renders decision on grievances between cantons, between the cantons and the Federal government, and on a wide range of civil suits. However, it cannot declare specific cases of legislation unconstitutional or even become involved in constitutional interpretations (Brooks 1927).

The short terms of office for all government positions show the concern of the Swiss for the possibility of corruption and usurpation of power with extended service. Very strict nepotism rules counteract any possibility of oligarchic collusion. Yet, the possibility for unlimited reelection indicates the desirability of continuity and efficiency in performance of services.

The preceding outline of Switzerland's history indicates that the transition from a low-level democratic complexity of direct, assembly-type decision-making in individual cantons (beginning ca. 1291) to the high-level complexity of the national Constitution of 1848 (revised in 1874), took two hundred years longer than did the development of the high-level complexity Constitution in England, and occurred about eighty years after the inauguration of the American Constitution in 1753. Repeated interference by cantonal special interests in the full development of the national Constitution suggest that the strong tradition of local governance in the cantons actually impeded and delayed the final steps in the development of a complex national constitution for the Swiss. In Switzerland's case, the complexity of its sociopolitical federal organization is so distinctive because it achieved a unification of just those heterogeneities of language, religion, and economic and cultural stratifications that has been reinforcing isolationalism in the cantons. A further indication of high-level complexity of Swiss democracy is the coexistence of three radically different forms of decision-making: the majoritarian elections of competing political parties for representation in the cantonal governments and the Federal Assembly and the majoritarian form of the law-making process in the Assembly itself, the consensual form in the National Council, and the direct expression of public opinion in frequent cantonal and nationwide referenda. It seems important that the consociational-interpretative role of the National Council balances the polarizing, centrifugal tendencies of the two majoritarian decision-making processes.

Recapitulating, there are several factors making the Swiss political system less prone than the American to polarization and to partisan blockage of the progress of legislation: the Swiss Federal elections are low-key and lack the tumultuous frenzy of the American party conventions; the public personality of the Swiss president is much less conspicuous than that of the American counterpart; all major ethnic-linguistic and religious elements of the Swiss population are represented in the executive apparatus; and the news media have minimal access to the actual executive decision-making process. In contrast to American elections, in the Swiss system the relatively high level of election contentiousness occurs on the cantonal level whereas in the American system the contentiousness is more conspicuous on the national (federal) level. Also, polarization of the political apparatus is diminished by deciding highly controversial issues through popular referenda, more readily available in the Swiss system than in the American system, where early on it was seen as a concession to mob rule. These characteristics are emphasized here because they promote the recognition and elaboration of specific spheres of common interest, of possibilities of joined legislation, of linking initially diverse approaches to governance by what can be seen as a sociopolitical equivalent of conceptual continuity. The following Chapter 6 deals with the rise of the sociopolitical simplifications of totalitarianism with the decline of the role of the individual in the conceptual continuity of society. A further step in the strengthening of conceptual continuity in democratic society is then considered in the concluding Chapter 7.

Chapter 6 The Threat to Sociopolitical Complexity: The Simplifying Sociopolitical Ideologies of Nationalism and Communism

This chapter deviates from the main theme of the book. It reemphasizes one of the main sources of the precariousness of sociopolitical complexity and is an extended preface to Chapter 7. The preceding chapters give an account of systems with increasing complexity. Democracy was considered as a sociopolitical system with a very high degree of complexity and multiple conceptual links between its basic elements. Certain societies chose, however, to evade high-level complexity and to achieve some simplification by reducing the number and variety of conceptual links in their social framework. These are the totalitarian forms of society that include nationalism, communism, and fundamentalism. That the societies that circumvent complexity become the most menacing threat to the preservation of humankind has become evident, but it is relevant to show in this chapter that at the initial stages of their development these societies were aiming toward high levels of complexity and only fortuitous circumstances determined their final detrimental fate.

Nationalism, examined first, appeared in the nineteenth century as an essential element of several sociopolitical settings (Kohn 1961, 1965;

Deutsch 1969; Hobsbawm 1990a). A brief survey of the nationalism that emerged from the French Revolution and its Napoleonic aftermath and the nationalism that evolved during the transformation of Germany into a unified nation and Germany's subsequent strife for European hegemony are included here. The relationship of complexity and simplification in the development of Marx's materialist dialectics is examined in the following section.

THE FRENCH REVOLUTION: PRELUDE TO
THE IDEOLOGICAL RIVALRIES OF
THE NINETEENTH CENTURY

The French Revolution of 1789 to 1815 dramatically heralded the conversion of the sociopolitical utopias of the Enlightenment into the concrete implementation of ideological simplifications of sociopolitical reality. It became a unique historical event that brought the main elements of sociopolitical reality into play. The storming of the Bastille can be seen as the first violent claim for social justice; the French National Assembly's Declaration of the Rights of Man and the Citizen as the beginning of constitutionalism; the execution of the king and the queen as the end of absolutist dynasties; the reign of terror as an exercise in authoritarianism; and the total mobilization of France's population as symbolizing the awakening of nationalist military aggressiveness.

Raised to the crests of the consecutive waves of the revolutionary process were some unique personalities who emerged and vanished in rapid succession, indicating the uncertain fate of the individual in a shifting social hierarchy. Although some of the simplifying general abstractions of the Enlightenment became incorporated in the flow of the revolutionary process at certain points, they seem to have become submerged by the particular circumstances that actually determined each step in the succession of events.[1] For these principal reasons, a reconsideration of the French Revolution may be helpful in examining the relevance of simplifying general abstractions or of the identification of specific conditions for the appraisal of sociopolitical systems. The main events in the revolutionary process and the elements that produced its dynamics are surveyed next.

The French Revolution was a sociopolitical event that has been intensively studied and documented (Furet 1968; Skocpol 1979; Rude 1988; Feher 1990; Stocpol and Kestenbaum 1990; Baker 1991; Chartier 1991). Just before its second centennial in 1989, an account of the full dimensions of the kaleidoscopic changes and the diversity of interpretations of the French Revolution led to the compilation of a *Critical Dictionary of the French Revolution* of more than a thousand pages (Furet and Ozouf 1989).

One thinks of the France of the Enlightenment as a site of high culture, one esteemed throughout all Europe. French was the language of high society, even at the court of Fredrick the Great of Prussia. France was prominent in literature and the arts: in philosophy, with the great *Encyclopedia* as its hallmark; and in science, with the mathematician Lagrange, the physicist Réaumur, and the chemist Lavoisier. France also produced the first paddle steamer and steam automobile, the Montgolfier balloons; and the precursors of the telegraph and telephone. The architecture of Versailles was imitated by the most prestigious courts of Europe. But the favorable image of French society camouflaged the conditions that led to the social eruption of the French Revolution. Symptoms of dissatisfaction with the absolutist regime had been manifest for about a century, and mounting discontent in all strata of the French population eventually led to a confrontation.

That this important Revolution took place in France during the last decade of the eighteenth century can be attributed to the coincidence of several sociopolitical conditions. Heavy tax burdens and poor harvests left a large part of the French peasantry at only marginal subsistence, and their awareness of the higher living standards of the expanding bourgeoisie enhanced their hopes for an improved livelihood. The bourgeoisie, in turn, was aiming for an end to aristocratic privileges, and sought a share in the legislative process and in the control of state finances. Finally, the nobility was dissatisfied because they were excluded from quasi-judicial and legislative bodies, the Parlements, that admitted only the king's appointees, and were opposed to any reform of the status quo of absolute monarchy. Meanwhile there was everywhere a general decline of prosperity, increasing food prices, and unemployment.

Facing these worsening conditions, King Louis XVI considered introducing reforms. When their implementation was delayed, in May 1789 representatives of the nobility (First Estate), clergy (Second Estate), and bourgeoisie—including the lower middle class (Third Estate) reconvened the Estates General, a constitutional legislative assembly that had been discontinued for more than a century because it had been interfering with the simplifying concentration of power in the hands of the absolute monarch. No immediate agreement was reached for a joint legislative function of the three estates in its first sessions. At this point, following the spirit of Abbe Sieyès's pamphlet *What Is the Third Estate?* that Estate assumed a leading role in establishing the National Assembly, a legislative body that eventually included all three estates. In August 1789, the

National Assembly issued a Declaration of the Rights of Man and the Citizen, which became the basis of the new constitution announced in 1791. The 1789 Declaration was an assurance of the basic rights of equality before the law, and of liberty, property, and security, and can be regarded as an extension of the rationalist tradition of the Enlightenment. Its tenets show the influence of Locke, Rousseau, and Montesquieu and also of the American Declaration of Independence of 1776.[2] It is noteworthy that Sieyès objected to those qualifications in the American Declaration limiting the power of the people. He envisaged that the French Declaration would give his countrymen an unqualified sovereignty. In the end it was one of those absolutist simplifications of governance that became a source of political oppression. Articles Three and Six of the French Declaration foreshadow this absolutist potential. Sieyès, who was one of the architects of the National Assembly, introduced here Rousseau's concepts of the "general will," and emphasized the idea of nationality at the same time. He proposed: "The nation exists before all, it is the origin of everything. Its will is always legal, it is the law itself"(O'Brien 1990, p. 48).[3] Omitting the individual, one could hardly think of a conceptually more discontinuous argument.

In actuality, the main aim of reform was still the change from an absolute monarchy into a constitutional one at this stage of the French revolutionary process. The king was to retain executive power, but the legislative and judicial functions of the government were to be assumed by the people. The Assembly was to implement the interests of all three estates. It was a step from the relative simplicity of absolutist governance serving interests of a small privileged group toward a more inclusive and complex constitutionalism. It was a complexity that needed compromise, as a step toward conceptual continuity.

It can be questioned whether this transition from absolute to constitutional monarchy would be classified as a revolutionary event in itself. The establishment of the principles of the Declaration and of the new constitution could have been taken as an extension of the ideas of the Enlightenment. What was to follow was but an overture that preceded the curtain-rising of the nineteenth century drama. The readily recognizable themes of full revolutionary change were manifest in events that paralleled the deliberations of the French Assembly.

In widespread areas among the peasants there existed an exaggerated fear of destitution and of dispossession by illegitimate force. As a result of their growing fears, in separate incidents, the peasants turned against the landowners and burned their possessions. Similarly, the lower strata of the urban societies, namely the laborers in the workshops and in other menial occupations, the

Sans-Culottes, revolted against an apparent threat of wage reductions in Paris, destroying property in the process.

At this point the storming of the Bastille in 1789, an event in the revolutionary process that seems totally detached from the precepts of the Enlightenment, became one of the prominent symbols of the French Revolution. Utopian, simplifying abstractions of the Enlightenment were replaced by the concreteness of specific action. Although that was among the predictable results of the continuing mismanagement of reform, several accidental factors came into play at crucial points in the course of events. The role of accidents is emphasized here as indeterminate variables and one of the main distinctions between ideal mechanistic and nonideal systems. Thus, the announcement of the dismissal of Jacques Necker, one of the king's ministers sympathetic to the cause of the Third Estate and the lower-middle class peasants and laborers, provoked the gathering of crowds in protest. Tollhouses used for the collection of import taxes on goods coming into Paris were burned down. Revolutionary orators called for armed interference. The gunsmiths' shops were ransacked, and large stores of arms were removed from the inadequately protected storage at the Hotel des Invalides. From there, the crowds converged onto the Bastille; there, negotiations on the withdrawal of guns from the Bastille battlements were in progress on the morning of July 14, 1789. When the negotiations faltered, the crowds attacked and penetrated into the Bastille over an accidentally unprotected drawbridge (Rude 1988, p. 55). The Bastille garrison opened fire on the attackers, but the troops were overwhelmed by the massive assault and had to surrender. The commandant and six members of the Garrison were executed on the spot. Irrational violence combined with accidental advantages determined the outcome of the first major event in the Revolution, and set the pattern for a large part of the overall revolutionary process. The rationalist, mechanistic precepts of the Enlightenment seemed to have blown away like shreds of paper in a howling storm.

The storming of the Bastille in July 1789 hastened deliberations in the Assembly and resulted in the Declaration of the Rights of Man and the Citizen, but it did not bring an end to the revolutionary drive. In reacting to worsening living conditions in Paris, the market-women of the capital marched to Versailles in October 1789, backed by the National Guard under Lafayette; there they demanded an adequate supply of bread and implementation of the measures that had been decreed by the National Assembly but was thwarted by the Crown. Although the king gave orders to meet the demands, the embittered crowd released its frustrations by occupying the royal palace, returning to Paris with the king and his family, who missed their chance to escape, and who were placed under surveillance in the Tuileries Palace. Other indicators of economic deterioration, especially devaluation of the currency and bankruptcy of the treasury, led to the confiscation of church properties, but further outbreaks of rebellion were suppressed by the Crown for the time being. Meanwhile, the National Assembly continued meeting through 1790, and prepared a definitive constitution.

The deliberations of the National Assembly during 1790 and into 1791 produced the definitive constitution, adopted in September 1791. It turned out to be a document in which certain abstract basic rights were asserted to be self-evident, and the specific concrete steps for the institution of a constitutional monarchy were considered in some detail. It reduced the privileges of royalty and nobility, attempted to define the relationship of church and state, and established a system for the election of deputies to the Assembly. Citizens who could pay a minimal tax chose the members of an electorate that in turn elected the deputies to the Assembly. The Assembly held full legislative power, with no checks or balances that could possibly interfere with its function. The king and his appointed ministers represented the executive branch of the government, but the judiciary was established as a separate entity. France became organized administratively into Departments as the largest subdivision, with districts, cantons, and communes as successively smaller administrative subunits. Elected officials were responsible for the administration of units at all levels. The arbitrary multiple taxes, mostly collected from the lower-middle class, were replaced by a more equitable tax system. As Rude (1988) points out, one of the aims of the constitution was protection against infringements by "royal despotism, aristocratic privileges, and popular licentiousness." At this time the monarchy was still in place, and Louis XVI, totally misinterpreting the situation, pronounced the Revolution over. What these details reveal is that the Constitution did not reflect Hobbes's or Rousseau's absolute utopias, but rather Rousseau's plans for the concrete constitutions of Poland and Corsica, and Montesquieu's and Burke's ideas of governance. The abstract simplifications both of absolutism and the Enlightenment were replaced for the moment by a concrete constitutional complexity as a step toward sociopolitical conceptual continuity (see Chapter 5). Stability was not restored, however, for a number of reasons.

One of the destabilizing factors was Louis XVI's failure to come to terms with the drive from absolutist to constitutional monarchy. He tried to resolve his predicament in the summer of 1791 by escaping from France after secret negotiations for asylum in Austria, France's potential enemy. His flight was intercepted at Varennes, and the royal family was brought back and put under house arrest in the Tuileries on suspicion of having negotiated with an adversary power. On July 4, 1792, a massive public demonstration in the Champ de Mars demanded the abdication of the king, and was met with military action: it led to firing on the crowd, wounding or killing many citizens. Inflamed by anti-royalist agitators, the populace invaded the royal quarters in the Tuileries twice in June and August of 1792, forcing the abdication of the king. Secret documents were discovered that showed Louis had conspired with the Austrian court to begin a military suppression of the revolutionary process. This confirmed the charge of high treason against him, and after a short trial Louis was sentenced to death and executed in January 1793. His wife, Marie Antoinette, an Austrian princess who had participated prominently in the conspiracy, was executed a few months later. An indication of the precariousness of the revolutionary process and of the absence of mechanistic certainty is that the execution of the king followed a vote of 361 to 360 in the Assembly, a one vote majority in favor of the death sentence. With the abolition of the monarchy the last vestiges of a formal stabilizing center of society became extinct. In trying to reassert the absolutist form of monarchy, the king had failed to grasp the opportunities for salvaging a constitutional monarchy during a time when the 1791 Constitution still envisaged the royal court as the executive branch of the government.

Other destabilizing conditions were the continuing shortages of food; the year 1792 opened with food riots in Paris, and counter-revolutionary uprisings flared in some other parts of France. Most ominously, the National Assembly had an inherent, low level of stability. It was an assembly of contending factions, lacking a mechanism of checks and balances to provide for both coexistence and competition. This deficiency permitted an escalation of a power struggle that was aiming at the total usurpation of control by one of the factions and the elimination of all the others. Initially both the power and the majority in the Assembly were held by a faction, the Girondists, who supported the monarchy and mainly promoted the interests of middle-class entrepreneurs. The Girondists were opposed by a minority of closely organized Assembly members, the Jacobins, who represented the underprivileged strata of the poor peasants and the wage-earning laborers, known as Sans-Culottes. The Girondists, using their power in the Assembly, promoted and lost a military gamble in risking war with Austria. In their place the Jacobins, under the leadership of Robespierre, took the reins.

As confirmed in the National Convention of 1792, the Constitution of 1791 had been modified in accordance with the Jacobin precepts and approached what might seem to us now to be a welfare state. Robespierre, in assuming full leadership of the Jacobin faction, delivered an outline of the aims of the revised Constitution in 1793. The members of the legislative body were to be chosen in a general election. For the first time, all male citizens became entitled to cast their votes free of voting fees. Work was assured for the greatest possible number of able-bodied persons, and some support was considered for the unemployed. Here was the closest approximation to the ideal promise of the Revolution. Military involvements, however, discussed below, created strains in France's social fabric and shifted its governance toward uncontrolled authoritarianism.

Seven months after the meeting of the National Convention (September 1792), which represented an apparently firm foundation of the Declaration of the Rights of Man and the Citizen, a Committee on Public Safety was organized under the auspices of the Jacobins. Soon after its establishment, it included the leading Jacobin personalities Robespierre, Saint Just, and Danton among its thirty-eight members. This Committee gradually assumed full control of most branches of the French governing body. It controlled the activities of the ministers, including shaping foreign policy; it appointed the generals of the army; it purged and directed local governments (Rude 1988, p. 101). A parallel but less powerful Committee of General Security assumed responsibility for police and internal security, the revolutionary tribunals, and local vigilance.

With dictatorial ruthlessness, the Committee of Public Safety under Robespierre purged the Jacobins' adversaries as well as the less radical members of the Jacobin group by executing them; Danton himself was among those sacrificed at this point (through 1793 and into 1794). In an attempt to counteract the disunity arising in the Committee, Robespierre made the mistake of withdrawing from the arena for a short time to redraw the aims of the Revolution more consistently. Returning to the scene of action, he delivered one of his great orations, one that presented for the last time his distorted interpretation of Rousseau's ideas, with the "Goddess of Reason" as a unifying principle. In his absence, however, his adversaries had conjured up damning accusations against him, so that Robespierre himself became yet another victim of the guillotine's blade in July of 1794. The complexity of democratic constitutionalism was thus abolished, and governance was assumed by oligarchic and authoritarian directorships.

Perhaps the main reason for the rapid deterioration of France's newly acquired democratic foundations was the sociopolitical stress of her military activity.[4] Trends toward war had started at two opposite points in the sociopolitical spectrum. The left-wing Girondist deputy Brissot was already agitating for war with Austria in 1792, and expected victory to further his own cause as well as that of his entire faction in the Assembly. He also anticipated that war would become a rallying point for the French population. From the opposite side, the royal establishment, led mainly by the king's wife Marie Antoinette, encouraged war with Austria because France's expected defeat would open an opportunity to reestablish absolute monarchy. Thus, France declared war on Austria in April 1792. The first campaign ended in defeat for France, and the Austro-Prussian Army advanced to Verdun, no more than 200 kilometers from Paris during the following months. Brissot had to withdraw from his leadership in the Assembly, and the Girondist ministers were dismissed from the court. The Jacobins, under Robespierre's leadership, had opposed Brissot's war plans, and therefore gained power in the Assembly. With the threat of the advancing Austro-Prussian enemy, Danton, then at the height of his powers before his execution in 1794, reassembled an army that decisively defeated the Austro-Prussian coalition at Valmy on September 20 of 1792.

The initial momentum of the French forces was greatly augmented by a total mobilization of the French population, and coincided with another declaration of war by France, this time against England and Holland (February 1793). Apart from the large-scale recruitment of able-bodied males into the army, large sections of the civilian population were inducted into the manufacture of war material, from ammunition and arms to the preparation of supplies and bandages; the old and adolescents alike became involved, in addition to the normal work force. The great military and domestic efforts proved successful, and in 1794 the army had cleared all of France of enemy troops.

The English siege of Toulon was raised through a campaign led by the young Napoleon Bonaparte, who thus achieved national prominence for the first time. Napoleon followed this by his victories in an Italian Campaign in 1796–97, one in Egypt in 1799, and in the Battle of Marengo in 1800. Now First Consul Napoleon dismissed the directorship; this was followed by his coronation as Emperor in 1804, and by his brilliant European campaign in 1805–07. The "Grand Army" became the rallying point for French self-esteem, the hallmark and the main pillar of what now styled itself "La Grande Nation."

The French Revolution was of such pivotal importance in the history of governance because it not only replaced one form of governance by a succession of transient radically different ones, but because it also created a fundamentally new sociopolitical system of values. In this process the French Revolution created nationalism as one of the most powerful elements of a sociopolitical reality. According to Nora (1989, p. 742): "It was the Revolution that gave the word "nation" its synergy and energy. It crystallized the word's three meanings: social (a body of citizens equal before the law), legal (constituent power as opposed to constituted power), and historical (a community of men united by continuity, by a common past and future)."

The concept of "nation" was realized and implemented when the delegates, elected by the population convened in what became known as the National Assembly at the very beginning of the revolutionary process. Subsequently, the Declaration of the Rights of Man and the Citizen established that "the principle of all sovereignty resides essentially in the Nation" (after Nora 1989, p. 745).

The Declaration of Rights, the motto *Liberté, Egalité, Fraternité,* the festive celebration of Bastille Day, and the Marseillaise, became symbols of the "powerful, unifying effect of the Revolution itself" (Nora 1989) that resulted in French nationalism. For the common man, nationalism provided an expansion of his self. It added a dimension to his existence and reinforced the feeling of shared power. The two conditions that can be recognized readily as the roots of French nationalism are the restructuring of French society and the ascent of France to the leading military position in Europe. Before the French Revolution, class barriers motivated the individual to identify with one's class and family rather than with the nation as a whole. Lowering of these barriers during the revolution, with the broadening of suffrage and the general military mobilization, shifted individual identities to the nation as a whole. The nation became the largest social unit with which ordinary citizens, irrespective of class, could associate their feelings of common heritage and common purpose and enhanced security.

The French example suggests some of the reasons why nationalism is regarded as an emotional experience. It has been defined as a "conglomerate of feeling" (O'Brien 1990) or "a state of mind, in which the supreme loyalty and its most essential element is a living and corporate will" (Kohn 1965, p. 9). The advance of nationalism in nineteenth century Europe and in the rest of the world in the twentieth century, suggests that the nationalist experience surpasses all others as the driving force in history. It was the emotional incandescence, as Nora (1989) calls it, that created the frightful ambivalence of the French Revolution: its solemn pledge of peace for the benefit of the world that was transmuted into a military aggressiveness that carried the messianic spirit of nationalism on the tips of bayonets and the intolerance toward the internal enemy, that generated the blood bath of the "terror" of the Revolution.

The question is before us of whether the different phases in the process of nationalization are qualitatively sufficiently different to permit us to assign levels of complexity or simplicity to them. One can suggest that the consolidation of the separate social classes with their different interests and the establishment of the French administrative districts represents a significant simplifying unification over the organizational discontinuities of the absolute monarchy that preceded it. From a different point of view, the 1791 Constitution of French governance is comparable to the schemes proposed by Montesquieu and Burke in its complexity. But the 1791 French Constitution lacked the components essential for its stability, namely, safeguards for the maintenance of the adequate division of power with a system of checks and balances. The absence of these stabilizing conditions created opportunities for a struggle between noncompromising political factions in which the power and effectiveness of individuals became a decisive factor in the course of the French Revolution; a dictatorial role was taken first by Robespierre, followed by the subsequent drift into Napoleon's absolutist imperial rule. The resulting restraints on individuals imposed by these dictatorships represent those far-reaching simplifications that are expected to give such systems only temporary viability.

Tragically, the message of the French Revolution was misread. It was not the nonviolently achieved complexity of the Constitution of 1791 and its concomitant approach toward conceptual continuity that was regarded as its major accomplishment. Instead, the French Revolution imparted a powerful double-faced legacy, first through its simplifying emotional, nationalistic symbolisms linked with its military aggressiveness, and the rejection of the rationalist principles that originated in the century preceding it. Second, it left the sharp-etched features of its social radicalism, with its disruptive force (conceptual discontinuity), that eventually replaced the idea of permanence of the preceding century with the idea of change as the more adequate representation of reality. These ideas are discussed further in the following sections on German nationalism and Marxist socialism.

THE GERMAN TRANSFORMATION OF NATIONALISM

The development of nationalism can occur much more slowly than it did in the French Revolution, and extend over many decades. A recent, widely referenced study of the development of nationalism in small European nations (Hroch 1968) has suggested that nationalism seems to arise in three phases. In its first phase it is initiated by a small group of intellectuals who are interested in language, culture in general, and history. This is followed by a second phase, characterized by a widening of patriotic activity, with large portions of the population adopting a nationalistic attitude. Nationalism may or may not enter a third phase, a state of aggressiveness and totalitarian repression. These three phases can also represent distinctive steps of simplifying restrictiveness. The development of German nationalism extends over a time span of about 150 years and clearly exemplifies the sequence of Hroch's three phases.

The idea of nationality emerged in Germany before the French Revolution, but it was detached from (and in some respects opposed to) the thinking of the Enlightenment. Further, German nationalism arose without any connection to the political actuality of the eighteenth-century Germany, at that time a conglomeration of more or less independent authoritarian principalities. During that period, an initial definition of nationality in general and of German nationality in particular was offered by Johann Gottfried von Herder (1744–1803).

One can hardly conceive of a life more remote from revolutionary or nationalistic violence than Herder's. Born in a small East Prussian town, the son of a pietistic and stern father who was a teacher, organist, and church warden, Herder received his higher education in Konigsberg and was directed mainly toward theology, broadened by extensive reading in science and philosophy. He held various positions as teacher, preacher, and private tutor, thus affording him opportunities for travel and study. Early on in life, Herder became deeply interested in language and literature leading to the publication of his first book, *On Recent German Literature: Fragments,* followed by his *Treatise on the Origin of Language;* the latter won the highest recognition of the Prussian Academy of Science. Later, he published the *Oldest Documents of the Human Race* and *Another Philosophy of History for the Formation of Humankind.* After several years of transient appointments, he married, and in 1776 established permanent residence in Weimar in an administrative position that was sponsored by Goethe (Nisbet 1979; Dietz 1980; Sauder 1987).

Seemingly in accordance with his wide range of learning, Herder's ideas about nationality developed against a cosmological background. He viewed nature as the source of greatest possible variety superimposed on an underlying unity and simplicity (Koepke 1987, p. 59).[5] The Earth's place lay between the other known planets, indicating a certain balance in the Universe. Its geographic and geologic structure is the result of cyclical upheavals that occur in accordance with some general laws. Herder endowed living organisms with a "vital force" which imparts their unique characteristics. Human beings are distinguished by their upright posture, the basis for finer perception. An upright posture also enables a human to be a creative artist and artisan and to become adaptable to different climates and different types of physical environments, and as a result gives a potential for freedom; this potential in turn leads to the development of reasoned choosing between alternative possibilities for action.

In Herder's view, human beings are slated to fulfill the requirements of humanitarianism via sympathy toward others and altruistic love (conceptual continuity) as a complement to sex. Concern for others is the basis for Herder's ideas of justice and truth, and for the understanding of the basic rule of not doing to others that which you do not want done to yourself (Koepke 1987, p. 61). The state of nature and its progress involves a chain of ascending forms and forces, but the actual characteristics of such forces remained undefined. Humanity is a link between the purely biological world, for example, of animals and the higher form of godlike humanity; thus a dualism is created in a human's existence (Koepke 1987, p. 62).

Herder develops the idea of nationality from a cultural point of view. For him, culture is the implementation and expression of the communal spirit that maintains a society. This culture is derived from the thoughts and feelings of the common people, and in this sense it is a popular culture in which folk song, folk poetry, and religion are expressions of the communal experience. A nation's uniqueness is therefore represented through its culture; it is not described by an abstract, general concept. For Herder the shared unique experience of a "folk"—the collective of the common man—is the equivalent of a nation. The extent of that national communality is bounded by its indigenous language. As Koepke points out (1987, p.9–10):

> Language and thought are indivisible. Language is a key to thinking, learning and life. Languages have a history; they change with the climate, the environment, and the historical conditions. . . . Communication cannot be achieved through translation, but only through familiarity with the other languages. Only in such a manner could diversity be overcome and a new unity of mankind approached. Such a future unity would therefore be based upon a diversity of national languages.

This is an opening statement for the development of pluralistic (and hence complex) nationalism in which each nationality evolves its specific characteristics for the benefit of the whole of humanity.[6] Language is the carrier of the communal spirit, a spirit that transcends specialization and self-interest; it is incompatible with social patterns like those envisioned by the French Enlightenment Philosophe Helvétius (1715–1771), among others.

Herder's thinking about society, state, and nation has been further clarified by comparison with Rousseau's ideas (Barnard 1988, pp. 285–321). Both thinkers contrast a hypothetical human in a State of Nature with humans in the civil state. In the former, humans exhibit their indigenous characteristics and are free of the artificialities and corruptions that come with transition into the civil state. Both Herder and Rousseau propose that humans in nature can satisfy their needs in a peaceful existence without competing with others (more or less violently) for the exaggerated wants that emerge in the civil state. Only in the State of Nature do humans reveal the "authentic essence of a people's inherent and distinctive character" that is the basis for a truly national culture. Thus nationality is the condition that most fully satisfies the individual's needs for social relatedness. The isolated individual or the smaller units of social integration are only incomplete parts of the totality of the national experience (Barnard 1988, p. 286).

Indigenous national characteristics were to be found in the common people, who led relatively unspoiled lives apart from and uncorrupted by the artificialities of the higher strata of society, the privileged elites in particular. There should be mutuality in the independence or interdependence of individuals in the social context; this precarious balance is maintained in the state of nature, but its continued maintenance is one of the main challenges for the civil state and for the establishment of sociopolitical conceptual continuity.

Human association should be the result of a "will" and not of "consent." The active participation of the individual citizen in the creation of an association is characterized by will. Consent is a passive conforming with a duty imposed by some external source. As Barnard points out, it is "a sense of self-identification through the conscious act of willing (that) binds one to what one has willed to appropriate as one's own." Both Rousseau and Herder considered sources other than rationality, in particular religion, in generating the social will, and both had reservations about the compatibility of Christianity with the tenets of the natural state "for having destroyed ancestral religions . . . that define the soul and the character of a people" (Barnard 1988, p. 289). Both suggested introducing a civil form of religion, that in its realization was perceived by its critics as a repugnant part of Rousseau's Social Contract. Perhaps most importantly for the later development of the nation was the agreement that the nation is the primary concept and the state the secondary superstructure in the association-forming process. The particular organization of a state is created (and can be altered or abolished) as an expression of the general will, with the nation yet remaining intact (Barnard, p. 291).[7]

Fundamental differences exist, however, between Rousseau's and Herder's positions. Rousseau based nationality—that is, the mechanism for setting up an association—on a purely political basis (the *Social Contract*) without any consideration of cultural factors. Herder's approach is the opposite; cultural factors are regarded as the decisive ingredients in the emergence of a nation. Accordingly, political or cultural nationalisms can be distinguished. Still, a balance between nature and culture must be maintained (Barnard 1988, p. 292). Most decisively, Rousseau envisioned the natural human as an isolated individual with at most animal language as means of communicating. Herder's natural human exists within a family with human language. The relatedness of individuals as an inherent part of his State of Nature is missing from the existence of Rousseau's natural savage (Barnard, p. 301). The nation is only the final extension of the relatedness or context between individuals. In Herder's view, the individual has a bond of family-relatedness along with his or her human language to express this relationship, a possible example of conceptual continuity. Language is "the embodiment of a people's thoughts and feelings" (Barnard 1988).

"But it is not any language or any society: human beings are creatures of a *particular* language and a *particular* society: they are what they are being born and brought up within a distinctive matrix of culture and language, the matrix of a people or Folk" (Barnard 1988, p. 302).

Language is one of the characteristics that sets human beings apart from animals, and Herder's religious outlook remained probably the strongest single motif in his representation of the world and humanity. Living in an era of Enlightenment with its emphasis on reason and the scientific approach, Herder tried to accommodate those trends in his world outlook.

Nationality as a manifestation of relatedness among the individuals of a society is a unifying, and hence simplifying, concept. Yet, Herder's stress on specificity as one of the characteristics of nationality and his conception of humanity as a commonwealth of diverse, mutually supportive nations introduced a higher level of complexity into the idea of nationalism. Further, the maintenance of a balance between the dualistic aspects of nationality adds another dimension to this complexity. The development of German nationalism is an example of an excessive shift toward the simplifying side of the balance.

Another important development in German nationalism is found in the work of Johann Gottlieb Fichte (1762–1814). Actually, Fichte's thoughts on nationalism, published as *Reden an die Deutsche Nation* (Speeches to the German Nation), comprise only a small, late part of his total output (Fichte 1846).

The larger part of his writings deal with an idealistic representation of human existence, one opposed to the empiricism of the Enlightenment, and his larger work shows how his ideas differed from the forms of nationalism developed later.

The foundations of Fichte's philosophy were established during his studies at the universities of Jena, Wittenberg, and Leipzig. During this time Fichte's thinking was initially influenced by Lessing's advocacy of freedom of thought and social tolerance, by Spinoza's pantheism, and by Kant's moral philosophy. After finishing his formal studies, Fichte held several tutorial positions, including a stay of several years at Zurich, where he published his first major work, *Critique of all Revelation.* In it, he made divine authority and the expression of the supernatural in human existence the equivalent of respect for duty. In this sense, God did not exist as a personal deity, but exists "in the recognition of a fundamental and sovereign moral dynamic in reality" (Tsanoff 1972, p. 193). Fichte's treatise found Kant's and Goethe's interest, leading to Fichte's appointment as professor of philosophy at the University of Jena (1792). There Fichte developed his philosophical system, an attempt to define the individual self and its relationship to society as a whole.

As Berlin suggests: "Fichte's concept of the self is not wholly clear: it cannot be the empirical self, which is subject to the causal necessitation of the material world, but an eternal, divine spirit outside time and space, of which empirical selves are but transient emanations; at other times Fichte seems ot speak of it as a super-personal self in which I am but an element—the Group—a culture, a nation, a Church" (after Berlin 1998, p. 570).

In a discussion of nationalism, it seems relevant that Fichte's individual self is not created merely by the inscription on its mind of concrete experiences. Each individual primarily expresses an idealistic or moral activity in form of attendance to duty. In Fichte's view, "It is the free, purely dutiful commitment of the will in ideal devotion. The conviction of duty is a challenge to any mere wants or empirical conditions. The dutiful will is the will of our true self, pursuing in unremitting endeavor an expanding goal. In the recognition of any course of action as my duty, I am also expressing a more fully realized recognition of myself" (Tsanoff 1972, p. 195).

The final aim of individual existence is the liberation of the self. A constitutional framework that suppresses that liberation is intolerable. Absolute monarchy or the rule of the pope are examples of this oppression. The liberated self is incompatible with totalitarianism (Schenkel 1933, p. 29; Fichte 1846, 7:1–256). In this sense Fichte's thinking is basically democratic, and this is confirmed in his rejection of hereditary privileges. However, Fichte considers constitutional monarchy an acceptable form of governance (Schenkel 1933, p. 7). In the constitutional state (Rechtsstaat), the state has no legislative role, but it does protect the development of the individual citizen into an autonomous personality. In this sense the constitutional state is the opposite of despotism (Schenkel 1933, p. 44).[8]

Fichte is concerned with the will toward realization of ideas (Schenkel 1933, p. 4). While Plato's individual remains in his realm of ideas, Fichte's ideas lead to action. He postulates that the individual, led by his ideas, will contribute to the moral development of society by his action; the individual is a free agent in the creation of the future. Therefore, it is Fichte's primary concern to educate the individual to spiritual freedom and moral responsibility (p. 5). In Fichte's system, this education is provided to any receptive individual irrespective of his social status. Any individual can become a spiritually independent agent and hence achieve political competence.[9]

Fichte deductively defines his abstract concept of the Common Will as the relationship of the individual to the greater social whole. A possible similarity between this concept and Rousseau's General Will remains undecided here. It is the task of constitutional law in its abstract sense to find the "Will that cannot be anything else but the Common Will" (Fichte 1846, 3:151; Schenkel 1933, p. 80). By insisting on the concrete realization of this abstraction, Fichte abandons the simplifying idea of the General Will and enters empirical complex sociopolitical reality (Schenkel 1933, p. 82). According to Hahn (1969) Fichte thought that the concrete and traditional structures of societies should be radically altered by abstract ideologists and social problems should be solved by practical adjustments of specific situations.

> The main function of the sociopolitical theory is the elimination of erroneous, confusing, incomprehensible and useless doctrines (Hahn 1969, p. 36).
>
> . . . On the one hand the sociopolitical theory represents only the abstract reflection of sociopolitical practice. On the other hand it is a guidepost for sociopolitical reality (Hahn 1969, p. 39; translated from German by the author).

Fichte's nonconventional interpretation of religion as well as his leanings toward republicanism led to his dismissal from the University (1799) and afterward he assumed only transient academic appointments. By then, Germany had entered the struggle against its conquest by Napoleon, and Fichte turned his efforts toward maintaining and enhancing the German national spirit. After Napoleon's ascendancy over Europe, Fichte abandoned Herder's pluralistic nationalism and substituted the singularity of German nationalism. In his *Speeches to the German Nation,* he attempted to maintain and boost the morale of his audience and to generate the feeling of the unique superiority of the German people. He exalted the primacy and continuity of the German people and the German language through history by introducing the epithets of *Urvolk* and *Ursprache* [the originating or primal people and language] (Glotz 1990, p. 74). The characteristics of the German people include all that is good, noble, and strong. New generations were to be rigorously educated with a discipline commensurate with the spirit of German superiority. The individual must always be ready to sacrifice his or her life for the community. Life has value only as source of the persistent and the eternal; but as such, individual lives have little if any significance, an attitude that led ultimately to the celebration of death and destruction.

The indivisible whole of a nation, a folk, cannot incorporate another ethnic group with a different language without far-reaching disturbance (Glotz 1990, p. 79). If foreigners join the German community, they have to remain "silent" until they have become completely acculturated, indistinguishable in every respect, and certainly linguistically indistinguishable from the German host community.

In its passage from Herder to Fichte, the idea of nationalism has clearly taken a distinctive turn. Herder dealt with nationalism in general, and included Germanic and Slavic peoples as well as other ethnic groups. The German people differ from the other nationalities but all of them are equals in their contributions to the well-being of humankind.

Fichte's philosophical doctrine (his *Science of Knowledge* and the derived *Science of Ethics*) asserted the free self-realization of the individual, the free choice of one's moral duties in accordance with one's free will; and it is this self-realization that determines the individual's relationship to society. The free self-realization of the individual in Fichte's ethic is not subject to nationality or to an authoritarian or totalitarian form of society. In this idealistic scheme the individual faces a reality of high complexity.

With the *Speeches to the German Nation* Fichte's idealistic representation of reality changes. The individual becomes subject to the discipline of an education that has as one of its main aims the indoctrination of loyalty to the German nation. There is now the beginning of a restriction of free self-realization of the individual. But the limitations now imposed on the self-realization of the individual by its commitment to German nationality were primarily a countermeasure against the stricter limitations that had been imposed by the French occupation. With his leanings toward republicanism and democracy, it is highly unlikely that Fichte himself would have condoned the use of nationalism as the basis for some form of totalitarian governance. Nevertheless, the path toward such a totalitarian simplification of reality became possible, even though an irreversible progress along this path was as yet in the distant future.

Fichte's *Speeches* shows a transition from Herder's academic initial phase in the development of nationalism to its second phase, the beginning of its popularization and, at the same time, a narrowing of its representation. This second phase was now intensified in the work of Ernst Moritz Arndt (1769–1860) and Friedrich Ludwig Jahn (1778–1852), aiding the surge toward awakening German nationalistic identity on a broad scale. Arndt was born on the Swedish-governed Baltic Island of Rugen, the son of an enterprising German farmer and a religious and imaginative mother. Arndt's youth was dominated by the romantic natural beauty of the island and by the conservative particularism of the Swedish culture of that time, under which alien nationals were free to maintain their cultural identities. After completing his secondary education, he studied theology at the University of Greifswald (1791–1793) and was ordained, but he abandoned a clerical career, and instead became a tutor. After traveling from 1798–1799, he outgrew his provincialism and became converted to German patriotism; he then began his academic career by taking a position in philosophy back at the University of Greifswald.

The role of nationalism in Arndt's world picture developed only during his last years at Greifswald and after he left his position there in 1806. Having become aware of the decadence in German society and facing the onslaught of the Napoleonic armies, while still at Greifswald, he began to work on his magnum opus, a historical-philosophical treatise *Geist der Zeit* (The Spirit of the Time), the first part of which appeared in 1806, followed by further parts in later years. In it, Arndt reviewed the history of mankind and compared the decline of earlier nations with the decline of the German nation of his time. From there he set out to exhort the German nation to rally, and to begin a social, political, and intellectual regeneration.

Arndt deplored the transition from the vitality of the Renaissance and the Reformation into what seemed to him the lifeless abstractions of the Enlightenment (Pundt 1935, pp. 50–51). Embracing romanticism, Arndt chided his contemporaries for not daring to venture into life.

Neither, says Arndt, are there any truly great historians. Why? Not because contemporary historians lack freedom or liberty but because they lack the simplicity, the energy and the imaginative talent of the old historians.

The modern period [Arndt's period] cannot cease interpreting and judging, it can no longer perceive the whole in the majesty of its organized unity, in which alone the moving world lives.

The world has come to a point where the sense of simplicity and virginal innocence, that which inspires its geniuses, has vanished. Courage, the impulse toward love and sacrifice, the quiet perception and contemplation of the beautiful and noble all have disappeared (Arndt, *Geist der Zeit, Part I,* 1908, pp. 41–42; Pundt 1935, p. 52).

These passages indicate a rejection of the analytical, scientific, and rational simplification of reality of the preceding century for the substitution of intuitive, and emotional experiences as a more meaningful representation of a simple form of reality.

But Arndt's attitude toward ethnic or national diversity was ambivalent. Although he still accepted a God-given cultural diversity of nations (Pundt 1935, p. 159), at the same time he harbored a chauvinistic disdain for the French, and despised the Turks (p. 56); yet he admired the English, and praised the Dutch. He was fervently opposed to any mixing of nationalities and in particular rejected any mixing with the Jews who had settled in Germany.

In principle, Arndt did not initially support the idea of racial superiority or inferiority (Pundt 1935, p. 159); he did see some merit in ethnic and racial differentiation and merely opposed the mixing of ethnic entities (p. 160). Later on Arndt emphasized even more than Fichte the maintenance of the German ethnicity and language through history, in an unadulterated form, as a prototype of a nation (*Urvolk*) and language (*Ursprache*) superior to other nations and languages: in particular, the French. The Germans came to be "the purest race and they spoke the purest language" (Kohn 1967, p. 255). Eventually, Arndt rejected any pluralistic universalism that included all nations as equals, and instead conceived a universalism under German leadership (Kohn 1967, p. 265).

Politically, just as Fichte, Arndt promoted the cause of liberty and rejected any form of despotism. He advocated constitutionalism, with parliamentary representation of all classes of society and equality of all before the law, and he rejected any divine right as a basis of governance. Within this constitutional framework, Arndt advocated a freedom of the press and of expression in general (*Geist der Zeit, Part 4,* 1908, pp. 40–62). He recognized that the dispersion of the German population among a multitude of principalities had deadened any German national consciousness, or feeling of national solidarity (p. 34). Arndt thought of Prussia as the potential leader in a drive toward German national unification and revitalization, and he believed it possible to combine the new nationalism exemplified in the French Revolution with some form of German constitutional governance (Kohn 1967, p. 253). His role model was the English constitution with its "responsible and free activity of the citizen as the bulwark of national strength" (p. 264). At the same time, Arndt maintained that the active human will be guided by some unknown forces that determine the fate of humankind (p. 255), and he is in favor of any war that enhances the vitality of nations and "stimulates men to heroic achievements" (Pundt 1935, p. 160).

Whereas Fichte's *Speeches* were so involved in style and concept that they could be understood, if at all, by only a small minority, Arndt's *Spirit of the Time* was widely read and understood. In addition, Arndt began publication of short pamphlets in which he addressed with relatively simple language, single and provocative nationalist themes such as The Rhine: Germany's River not Germany's Border. Some of these pamphlets appeared in several editions and were sold by the thousands. With Arndt, the promotion of German nationalism emerged from the safety of the ivory tower and entered the public arena. There, as soon as Napoleon's armies were repulsed, antiliberal forces gathered strength as part of repressive European policies masterminded by Austria's Chancellor Metternich. Arndt's stand against despotism and any other form of absolutism in some of the German principalities, along with his advocacy of constitutionalism and freedom of the press, led to his arrest in 1820 and his dismissal from the University of Bonn. Although he was not put on trial, he remained under surveillance for twenty years, and was not reinstated until 1840. Thus, Arndt played an important role as an activist: with the popularity of his writing German nationalism had clearly moved into the second phase of its development. The simplifying singularity of German nationalism came to predominate public discourse and Herder's pluralistic nationalism faded to the academic background; but Arndt's nationalism still protected and promoted the freedom of the individual, with at most a shift in emphasis from cosmopolitanism to German communality.

The effectiveness of Arndt's literary and political activism was reinforced by Friedrich Ludwig Jahn through his main work *Deutsches Volkstum* (German Nationhood; 1991), in which he reemphasized the ethnic unit as the foundation of the national state. Moreover, he linked a call for the study of the German folk, the German ethnic community, to the creation of three organizations dedicated to disciplined national service: the military free corps of patriotic volunteers; the gymnastic associations for the training of patriotic fighters; and student fraternities imbued with nationalistic enthusiasm (Kohn 1967, p. 269). The direct experience of nationalism was broadly spread over all social strata of the German population through these three organizations.

In his *Deutsches Volkstum,* these organizations were concrete realizations of Jahn's concentration on the ethnic unit as the basis for the National State: the people's state, comprising a single nationality. Jahn rejected the idea of a world community and of a world language (Kohn 1967, p. 275) with the strive for unity limited to boundaries of ethnicity, and the ethnic group defined by its historical tradition and its common language. In essence, Jahn admonished the German people, as had Arndt, to emancipate themselves from the foreign customs that had been adopted under the French occupation, and to revitalize German character and ambition. With this concrete implementation of academic ideas about nationality, Germany moved a decisive step closer toward establishing a precedent for the organization of radical nationalist paramilitary forces a hundred years later during the rise of Hitler's National Socialism.

It is important here to contemplate the relationship of German nationalism to two other, co-existing sociocultural trends, romanticism and cosmopolitanism, that affect the individual perception of the levels of sociopolitical simplicity or complexity.

In its romantic attitude, the individual senses the rich diversity and uniqueness of personal experience. This reality is complex, and it cannot be grasped by unifying rational concepts, but only by equally diverse and subjective experiences. The German romantics found a literary expression of their reality in the dramas of Friedrich von Schiller in his portrayals of the struggle of individuals against any form of oppressive authority, as in the *Wallenstein* trilogy, the *Maid of Orleans* or *Wilhelm Tell.* Goethe represented the romantic spirit in his early dramas such as the *Gotz von Berlichingen,* where he suggested the disdain of the individual for the restrictions of a bureaucratic society. His *Sorrows of Young Werther* became a sensational account of the love-sickness and suicide of a German youth, a work that aroused the compassion of readers all over Europe. Some Romantic writers glorified the adventures of the medieval knights; others explored the lightness and darkness of human nature (Eichendorff, Novalis); and yet others expressed a yearning for the simplicity of a utopian past. In what Whitehead (1925) calls the Romantic Revolt against Science, the romantic rejected the simplifications of reality in the form of scientific abstractions.

In cosmopolitanism (universalism) the individual identifies with a multinational expanse of humanity (conceptual continuity). Napoleon's aim appears to have been a cosmopolitan unification of Europe under French dominance, a unification widely accepted among certain strata of German society. In his post-Romantic period, Goethe welcomed the Napoleonic cosmopolitanism: he saw Germany at the center of Europe, open to influences from both the West and the East. In Kohn's words, "Germany becoming truly the heart of Europe, generous and full of sympathy, a guardian of peace and equilibrium, constitutionally unable and psychologically unwilling to centralize its forces for aggression and expanding purposes" (Kohn 1967, p. 169). The romantics were opposed to the mature Goethe, whose goal of education was the well-rounded harmonious man (was that an education to understand the complexity of cosmopolitanism?), and to "the personality which willingly subordinates itself to bonding forms and to the obligations of universal laws which rejoice in measure, symmetry and proportion" (p. 169).

It becomes apparent that the levels of the simplicity or complexity of reality are readily discerned only when qualitative differences are apparent. It seemed justified to regard the highly restrictive Spartan society as the simpler model, and Athens as the more complex model of sociopolitical reality. Similarly, the mechanistically perceived utopias of the Enlightenment seemed to represent the simple, and the constitutional monarchy of England the more complex forms of human existence. There are, however, certain characteristics of nationalistic simplification that may be used as indices, such as freedom of the press. These were abolished in all totalitarian forms of nationalism and communism, and are considered later on in this chapter as distinctively restrictive, simplifying forms of sociopolitical reality.

If the individual is considered as social unit, the steps to nationalism and cosmopolitanism can be perceived as distinct increases in complexity. If with each step up in this organizational hierarchy, some freedom of independent individual action, and hence of complexity, has to be sacrificed, a quantitative assessment of complexity becomes necessary in order to decide whether the advance in this hierarchy entails an increase or decrease in the experience of complexity for the individual. As a further reservation in evaluating levels of complexity, we have to consider that romanticism and cosmopolitanism prevailed only in small elites of German society, whereas nationalism encompassed a significant and rapidly increasing part of Germany's population. Both the expansion of individual experience to the richness and diversity of the romantic self or its expansion to the inclusiveness of the cosmopolitan nonself requires a relatively advanced level of literacy and general education. In contrast, the experience of nationalism can be obtained in organized social gatherings like nationalist song clubs, the gymnastic groups of Jahn or in military and para-military service.

As pointed out earlier (see the preceding section on the French Revolution), in this context nationalism can be regarded as a powerful emotional extension of the individual self, that can generate a dual experience by adding the com-munal self to the individual self. By adding a dimension to individual experi-ence, nationalism increases its complexity; but by imposing restrictive de-mands, such as service in the armed forces, it reduces the complexity of individual experience. Depending on its extent of restrictive measures, nation-alism can create a net increase or decrease in the extent of individual experience and hence of complexity in individual existence. English nationalism, with a minimal extent of restrictions, probably adds to the total complexity of individ-ual existence. French nationalism under the Jacobin dictatorship and the subse-quent directorates probably diminished individual complexity with its exten-sive restrictive measures.

Just as Germany was led into its initial simplifying unification by Bismark's domineering personality in the late nineteenth and early twentieth centuries, so it was Hitler who enacted and symbolized the ultimate simplifying realization of Germany's nationalism. Hitler's life was as much the product of his time just as he, in turn, shaped European destiny through his role in the last phase of German nationalism, with the record of his past closely interwoven with two crucial decades of history.

Hitler was born in 1889 in Braunau, a provincial Austrian town with a predominantly rural character. Growing up without a father and with a doting mother, he dropped out of school and became a drifter. He ended up in Vienna, eking out a marginal existence while attempting to follow a career as an artist. He failed to establish himself first as a painter and later as an architect. He found an outlet for his frustrations in the anti-Semitism rife in Vienna at that time (Schorske 1979). The intensity of this anti-Semitism and the incipient dissolution of the Austro-Hungarian monarchy, with the Slavic minorities threatening to overwhelm the German nucleus of the empire, seems to have left deep imprints on Hitler's personality. He became convinced that a multi-ethnic, constitutional government was not viable, and that foreign ethnic groups were endangering the German ethnic identity.

When Hitler turned twenty-four, the age for induction into the Austrian Army (1913), he evaded conscription by crossing the border into Germany, and took up residence in Munich. When Germany entered the war against the

Balkans and soon thereafter against the Allies, Hitler enthusiastically volunteered and became a messenger at the front, remaining in the army until 1920. He was never promoted beyond the rank of corporal. Nevertheless, according to Herzstein (1974, p. 22), his service during the war became for Hitler perhaps the first and most meaningful experience of his existence. The German armistice was a terrible psychological blow for him that became a further determinant of the rest of his life. Hitler saw the defeat not as the result of military incompetence and of exhaustion due to the blockade of Germany by the Allies, or as the result of the Allied superiority, but as the result of internal conspiracies led by Bolsheviks, Social Democrats, and Jews (Herzstein 1974, p. 24).

In the aftermath of the First World War, widespread but localized antimilitarist violence and general chaotic conditions led to the abdication of the German emperor and by 1919 the Weimar constitution, based on the French Declaration of the Rights of Man and the American form of government, became the basis for the German Constitution. This consisted of a federal system with considerable power left to the individual states, but it provided for proportional representation, giving radical splinter groups access to seats in parliament. Fatefully, the constitution decreed that in times of crisis the chancellor could assume dictatorial powers and issue orders, which on consent of the President, would become law. In avoiding a full-scale revolution, Germany became a prevalently middle-class republic that left the strength of the industrial and agricultural magnates essentially unaffected. For example, an attempt to establish a communist regime in Bavaria was crushed by a brutal paramilitary organization, the *Freikorps,* one of many extremist groups that had sprung up as associations of former veterans, for whom the war had been the great adventure of their lives, and who could not adjust to an ordinary routine of existence.

After having retained a position in the post-war army for some time, Hitler became an informant on the various nationalist groups being organized at that time. As one of those assignments, he attended a meeting of what became known as the German Workingmen's Party. Hitler eventually joined that party and became a member of its steering committee, while continuing his service in the army at the same time, primarily as educational lecturer to groups of former soldiers. His oratorical skill found the attention of the German Workingmen's Party, and he was made a recruiting officer. The name of the party was changed at that time to National Socialist German Workingmen's Party, with an aim of national rehabilitation by counteracting the harshness of the Treaty of Versailles. Having advanced into the higher ranks of the party, Hitler prematurely and unsuccessfully attempted a takeover of the Bavarian government in 1921, was arrested and jailed until 1924.

Hitler's imprisonment signaled the beginning of a short remission in the radicalization of German politics. The intolerable burden of the reparations imposed on Germany by the Treaty of Versailles was mitigated by the Dawes and Young Plans and in large areas of its economy, Germany became competitive in the world market. The political leaders of France and Germany, Briand and Streseman, were striving for a rapprochement between the two countries, and proposed a scheme for a federation of European countries. In addition there existed an independent pan-European movement that advocated a United States of Europe. From 1924 to 1928, the number of votes for the democratic majority socialist parties increased from six million to nine million, whereas the votes for Hitler's party (the National Socialists) dropped from about two million to eight hundred thousand. During the same period there was an unparalleled outburst of cultural creativity, covering all forms of art and science. Clashes between political extremist groups—Nazis and communists—happening especially during weekends were taken for granted as unavoidable emotional outlets for social outcasts, and a feeling of promise and security and acceptance of sociopolitical complexity prevailed. Germany's economic development, however, was supported by loans, mostly from American banks. After the Stock Market Crash of 1929, those loans were withdrawn; worldwide unemployment and widespread unrest followed, particularly in Germany, and this became the opening for Hitler's ascent. Using his skill for drama and emotional persuasion, he promised to satisfy not only the dire material needs of a large part of the population, but also their irrational longings; for the people had become alienated from the complexities of modern, democratic existence. As a consequence, support for the National Socialist party increased sharply; it reached parliamentary equality with the Social Democrats in 1930, and a more than 2 to 1 majority in 1933. In that year Hitler assumed the chancellorship by political manipulation, and substituted a totalitarian regime. As the only party, the National Socialists were unanimously elected in 1934 by 40 million voters.

At that point, Hitler introduced into German nationalism the simplifications which characterized its final phase: There was one superior race, the Aryan race, with the Germans as its most ideal representatives. This superiority was not to be jeopardized by admixture of inferior races, in particular by Jews. This danger was to be most effectively eliminated by an absolute solution, the extermination of all Jews. Other ethnically inferior populations, such as the

Slavs and the Latins, had to be kept in subjugation. Hitler treated dissension in the National Socialist party itself with the same radical simplification as external threats, by murdering two high-ranking members of the party who had been promoting a socialist line in the party program. All cultural activities were streamlined to serve the national socialistic cause, and the press was strictly censored.

Intense feelings of superiority were the basis for an oversimplification of reality that led Hitler to make risky moves. Some of these moves turned out successfully, such as his occupation of the Rhineland and of Austria (1938). The greatest triumph of Hitler's superiority complex was the surrender of the Czech fortifications by the infamous Munich agreement between Germany, Britain, France, and Italy in 1938. The same oversimplifying superiority complex betrayed Hitler during the Second World War, when he overextended the advances of the German Army in Africa and in Russia. The defeats of the German army in these campaigns became the turning points of Hitler's fortunes, and heralded the end of this phase of the nationalistic simplification of reality.

Until the end of the eighteenth century, nationalism was an exception in social organization. In most of the Western realm, the prevalent forms of governance were large multiethnic empires. Marriages among the ruling families and military conquests shifted the boundaries of these domains with disregard of the ethnicities of their inhabitants. In other instances, for example in eighteenth-century Germany, a relatively homogenous ethnic population was divided into numerous administratively separated principalities. Up to the modern era, the individual had hardly any conception of social units larger than those of the daily experiences of family, village, or township. In the mostly rural populations of that time, individuals and their small communities were too isolated from one another to generate an awareness of nationality. The upper social classes, the clergy, the members of the aristocracy, and the academics were aware of the history of antiquity and of the great medieval empires, but this did not promote thinking along nationalistic lines. The great religious schisms preceding the Enlightenment followed political, but not strictly nationalistic, lines of separation.

The rise of an influential middle class and the accompanying increase in urbanization and literacy led to the recognition and strengthening of a common tradition, a common language, and common economic interests. This became a basis for the development of the concept of nationality. Depending on its region of origin, nationalism developed in three distinctive forms (Kohn 1967, p. 2). A nationalism of considerable tolerance incorporating traditional and modern elements developed in England, Holland, and the Scandinavian countries (Hroch 1968; Mitchison 1980; Samuel 1989; Butt 1980. There nationalism did not diminish the complexity of the internal structure of those countries, and imposed little or no restraint on individual freedoms; this enhanced the uniqueness of the respective nations and enriched Western culture as a whole. This can be regarded as a cultural, permissive, and nonrestraining form of nationalism.

Other forms of nationalism led to far-reaching restrictions of individual freedom and the domination of other nations (restrictive, political nationalism). The development of this second form of nationalism is exemplified by the French Revolution, which began by creating a nation-state with equality and liberty of the citizen as its basic tenets. However, within two years a cultural and permissive nationalism based on enlightened reasoning gave way to the fanaticism of the revolutionary terror and the military, authoritarian nationalism of Napoleon. Similarly, German nationalism began at the end of the eighteenth century from inclusive, cultural roots (Herder's national pluralism). It became increasingly restrictive as a countermove against both the French domination under Napoleon, and against the indigenous cosmopolitan culture prevailing in Germany at the end of the eighteenth and the beginning of the nineteenth centuries. The development of German nationalism became a long, drawn-out, discontinuous process eventually culminating in the monstrosity known as Hitler's National Socialism.

The nationalisms dealt with in this chapter emerge from completely different sociopolitical backgrounds. In France, nationalism appeared as an intrinsic element of revolutionary change, after the abolition of the monarchy. In Germany, nationalism began with a vision of its cultural uniqueness and its potential for contributing to the benefit of all humanity. Nevertheless, the two nationalisms share certain fundamental tenets. They relinquish through its repressive misuse (Robespierre) or its outright renunciation (Herder) the abstract, in particular the mechanistic, reasoning of the Enlightenment and instead substitute the concrete emotional experience of communal relatedness and security.

At their inception, the two forms of emerging nationalisms were seen as coexisting with many other unique nationalisms. The French Constitution of 1792 pledged that the French nation would maintain a peaceful harmony with its neighbors, and Herder wrote of each nation using its unique skills for the benefit of all other nations. These nationalisms, in their early phase of development, were complex and pluralistic, and they emphasize the unique differences between nations, rather than giving a single abstract definition of a general form of the state as the only representative sociopolitical reality.

The pluralistic complexity of nationalism is linked to the unifying experience of communality by the individual. Therefore, it is perhaps not just the pluralism of nationalism itself, but the maintenance of a balance between the independent individual, the individual as a unit in the communal collective of a nationality, and of the nationality in its coexistence with other distinctive nationalities that represents a high level of sociopolitical complexity. If this high level of sociopolitical complexity is not maintained, the system collapses into a condition in which individual and nationalistic pluralisms are obliterated via the takeover of authoritarian or totalitarian forms of governance. This happened in both French and German nationalism. Such a dilemma was avoided, beginning with the English form of governance, with its controlled checks and balances and division of power. It was developed with maximal safeguards against totalitarian takeover by the American Constitution and its amendments, thus giving it sufficient flexibility to overcome adverse conditions as discussed in the preceding chapter.

CHANGE AS THE BASIS OF SOCIOPOLITICAL REALITY: THE DIALECTIC INTERPRETATION OF HISTORY IN MARXIST SOCIALISM

Toward the end of the Enlightenment, there emerged a dynamic conception of reality in which historical change acquired primary importance. This reality appeared in Georg Wilhelm Friedrich Hegel's new form of dialectics, and it became one of the roots of the sociopolitical ideas of Marx and Engels.

In Hegel's (1770–1831) dialectic, divergent or even opposite forms of reality, called "thesis" and "antithesis," become related in a "synthesis" as "fluid categories that constantly amend each other" (H. B. Acton 1972, p. 441). This changing and evolving system of opposites is particularly relevant in his interpretation of history. For Hegel, each stage in history is thought to be essential for the development of the next stage; each stage is justified as preparation for its union with its opposite as the subsequent step in the historical advance. In this sense, the dialectical progression is a deterministic process, although the conditions determining the transition from one dialectical phase to the next are not necessarily known in each instance. Ultimately, and especially in its historical form, the dialectical process is as unalterable in its dynamic advance as the orderly repetitiveness of planetary motion. In establishing his dialectic, Hegel referred to numerous concrete examples from history and nature; subsuming a wide range of apparently heterogeneous events and phenomena under one general concept gave rise to a corresponding simplification of reality.

Hegel distinguished between two types of knowledge. One of these, called "reasoning," is the basis of his dialectic. Reasoning consists of a purely abstract conceptualization without reference to experience or observed data. For example, Hegel was searching boldly for numerical relationships between the planetary orbits that would demonstrate an a priori cosmological principle of greater generality than gravitational attraction. Newtonian mechanics, based on observed data, represented a second, lower form of knowledge, which Hegel called "understanding." The distinction was derived from Immanuel Kant's corresponding categories of Vernunft und Verstand.

Also according to Hegel, reality could be represented only by the characteristics of the system as a whole, and not by the relationships among isolated parts. Any analytical approach that might attempt to define the state of a system, or its changes, by accounts of its parts will remain incomplete: a position anticipating what was later to become known as holism.

Hegel taught for some time at the University of Berlin, and for many years after his death, his dialectic remained a subject of much discussion there. During his studies at Berlin (1836–1837), this provided Karl Marx (1818–1883) with a basis for developing his own version of dialectical thinking. The elaboration of dialectical thinking into the foundation of the communist doctrine was the result of a lifelong collaboration of two incongruous personalities, Karl Marx and Friedrich Engels, bound to each other by their common aim of leading humankind toward an ideal, classless society. Marx's main contribution was the establishment of the socioeconomic basis of communism. Engels worked on relating dialectical thinking to the advances of nineteenth century science.

The Marx-Engels system is of special importance here because it is probably the outstanding example of how inadequate abstract simplification is for representing the concrete complexity of sociopolitical systems. Marxist dialectics, which were based on tenuous assumptions, produced a wide range of unfulfilled predictions. By creating a prospect for a total solution of social problems, and an apparent certainty of its promise for the elimination of human suffering, the new dialectic was adopted as faith by an ardent, aggressive, and uncritical following. The eventual impact of this system on a large part of humankind was out of all proportion to the invalidity of its premises.

Although the term "dialectic" is used as the common heading for both its Hegelian and Marxist forms, it covers two divergent meanings. In the idealistic

Hegelian interpretation, the dialectical process has an abstract, conceptual origin in the human mind that then assumes concreteness, for example, as a phase of historical development.

In contrast to Hegel's idealistic dialectic, Marx's materialistic interpretation finds the dialectical process inherent in nature and in the sociopolitical process; it is then secondarily reconstructed into an abstract, conceptual scheme.

Disregarding for now the discrepancy between the two forms of dialectic, it is important that both types were thought to be more inclusive, broader generalizations that would lead to more fundamental simplifications of reality than did the concepts of the Enlightenment. Thus, Engels criticized the scientific achievements of past centuries, in particular Newtonian mechanics, as purely inductive collections of facts (see Hook 1975, pp. 64–85). Engels clearly stated the full contrast between the views of scientific reality of the eighteenth and nineteenth centuries in the introduction to his *Dialectics of Nature* , when he wrote:

> What especially characterizes this period [seventeenth and eighteenth centuries] is the elaboration of a peculiar general outlook, in which the central point is the view of the *absolute immutability of nature*. In whatever way nature itself might have come into being, once present it remained as it was as long as it continued to exist. The planets and their satellites, once set into motion by the mysterious "first impulse" circled on and on in their predestined ellipses for all eternity, or at any rate until the end of all things. (emphasis added).

The "immutability" included all terrestrial features: oceans, mountains, rivers, and plant and animal species. "All change, all development in Nature was denied" (Engels 1940, p. 6). Even more revealing is Engels's comparison between understanding of reality in antiquity and that during the Enlightenment:

> High as the natural science of the first half of the eighteenth century stood above Greek antiquity in knowledge and even in the sifting of its material, it stood just as deeply below Greek antiquity in the theoretical mastery of this material, in the general outlook on nature. For the Greek philosopher the world was essentially something that had developed, that had come into being. For the natural scientist of the period that we are dealing with it was something ossified, something immutable, and for most of them something that had been created at one stroke. Science was still deeply enmeshed in theology. Everywhere it sought and found its ultimate resort in an impulse from outside that was not to be explained from Nature itself. Even if attraction, by Newton pompously baptized as "universal gravitation" was conceived as an essential property of matter, whence comes the unexplained tangential force which first gives rise to the orbits of the planets? (Engels 1940, pp. 6–7)

Engel's main objection to Newtonian mechanics was the absence of attempts to come to an understanding of the origin and the evolution of the observed phenomena, from the planetary orbits to species of plants and animals. He welcomed, therefore, the Kant-Laplace theory (1775) of the transformation of galactic nebulae into planetary systems (Hook 1975, p. 9). What seemed of primary importance to Engels was the coming into being and the passing away of natural systems as a main (dialectical) characteristic. He found support for this in the observations and interpretations of geological phenomena of his day, which indicated that the earth and the life on it had an evolutionary history. Engels conceived of a remarkably adequate dynamic picture of geogenesis, involving the cooling of the earth with formation of the atmosphere and the oceans, the formation of chemical compounds from chemical elements, and the appearance of life. Of equal importance in giving change a strong scientific basis was his recognition of the interconvertibility of the different forms of energy into one another.

Engels subsumed the different forms of change under a single "Law of Transformation." This law states that the quality of a body cannot be changed without addition or subtraction of matter or motion (Engels's equivalent of energy) or both, as exemplified in the transformations of energy and of chemical reactions (*Dialectics of Nature,*1940, p. 27). Similarly, conversions between the solid-liquid-gaseous states are qualitative changes, due to quantitative addition of heat or pressure. In chemistry, change is due to addition or removal of atoms from molecules.

It is apparent that this Law of Transformation cannot be regarded as a definition of physical, biological, or sociopolitical change, and it is hardly more advanced than any of the major "theories" of physical reality proposed during antiquity. It shows that Engels, in his disdain for Newtonian mechanics, completely missed its strength, namely the establishment of quantitative relationships between the variables in a system and the construction of what approaches conceptual continuity between the related elements. To propose that dialectical change was not just a metaphorical description of nature and of society, but was in fact the most comprehensive system of thought for representing all forms of reality that are in a way superior to traditional kinds of science, indicated Engels's total lack of understanding of science and its limitations. Despite its vagueness, the assertion of its superiority by Marx and Engels established dialectics as the main approach toward a solution of the fundamental social problems of the time as indicated in their *Communist Manifesto* (1848).

The *Manifesto* postulated that all of history is the result of the materialist dialectics of the struggle between social classes. In the course of this process, the bourgeoisie eliminated feudalism and other patriarchal relationships. At the time the *Manifesto* was written, the contending classes were the bourgeoisie and the proletariat. The bourgeoisie was blamed for the greed of free enterprise that led to the exploitation of the working people. For Marx and Engels, bourgeois economy necessarily created crises by the overproduction of commodities, with periodic disorganization of production and labor's resulting unemployment and suffering. The introduction of machinery and the division of labor deprived labor of all satisfaction in the production process itself, and led to an alienation of laborers from their work. The *Manifesto* proposed that with the advance of industrialization the labor force would become a majority of the population. At the same time, capitalist competition would lead to a concentration of the means of production in fewer and fewer hands. This then establishes the conditions for the takeover of the means of production by the proletariat itself, by allowing labor, as the new majority, to prevail in a democratic process and, only if this is denied, by revolution. Leading the takeover is the Communist Party "with its clear understanding of the line of march, the conditions and the ultimate general rules of the proletarian movement" (Padover 1971, p. 91).

The position taken in the *Communist Manifesto* was based on Marx's introduction of socioeconomic conditions as the main determinants of human history. Its view of reality is a radical departure from earlier thought. During the Enlightenment it was thought that the human mind, and with it, consciousness, forms as the result of sensory experience of its environment by an individual self. Secondarily, the individual acts upon its social milieu according to its own consciousness. In Hegelian systems, consciousness arises as an autonomous process of the human mind. Here again it is the individual's consciousness that affects its social environment. For Marx, it is the mode of economic productivity that is the primary given fact determining not only the state of society but also, secondarily, the consciousness of the individuals within society. Marx emphasized this position in a well-known passage from his *Critique of Political Economy* (Marx 1970, p. 20).

My inquiry led me to the conclusion that neither legal relations nor political forms could be comprehended either by themselves or on the basis of a so-called general development of the human mind, but that on the contrary they originate in the material conditions of life, the totality of which Hegel, following the example of English and French thinkers of the eighteenth century, embraces within the term civil society; that the anatomy of this civil society, however, has to be sought in political economy.

In the social production of their existence, men inevitably enter into definite relations, which are independent of their will, namely relations of production appropriate to a given stage in the development of their material forces of production. The totality of these relations of production constitutes the economic structure of society, the real foundation, on which arises a legal and political superstructure and to which correspond definite forms of social consciousness. The mode of production of material life conditions the general process of social, political and intellectual life. It is not the consciousness of men that determines their existence, but their social existence that determines their consciousness.

At this point a general dialectic became a special dialectic of sociopolitical economy. The simplification of the diversity of natural processes by a general dialectic was now supplemented by a simplification of historical diversity, via a single deterministic interpretation of socioeconomy that superseded the uncertainties of the "invisible hand" directing Adam Smith's market.

The clue to Marx's socioeconomy is the introduction of the notion that the amount of labor needed for the production of a certain commodity is the equivalent of its value. In preindustrial society the products of labor were of use to the laborer only. Carpenters of that time might make furniture that they could use in their own homes; Marx would consider this as "use value." Most of the laborer's work would produce the product—furniture—that would be used as exchange for goods needed but not produced by the laborer, such as food and clothing. The exchange of the furniture made for a certain amount of other goods is what Marx regards as the furniture's "exchange value." As the crucial point in Marx's economic system, the exchange value of any commodity is determined by the average amount of labor needed to produce that commodity. The minimum wage paid in the form of currency is equivalent to labor time and exchange value.

Marx thought he had convincingly demonstrated the basic tenet of his political economy: that the workers producing the commodities receive a compensation for their labors just sufficient to keep them alive at a minimal subsistence level. The value at which the commodity is sold is greater than the wage paid the worker; the difference constitutes the profit of the bourgeois owner of the means of production, and constitutes what Marx called the "surplus value." The introduction of a defined numerical magnitude (average labor time) as an equivalent for the much-less-defined qualities of exchange value and surplus value seemed to replace the complex arbitrariness of economic relationships with a consistent, general, quantitative simplifying system. Marx's numerical magnitude for average labor time provided a conceptual continuity between commodity, exchange value, and surplus value. According to Marxist doctrine, only the ownership of the means of production by the proletariat could effect an equitable adjustment between surplus value and wages. His redefinition of what appeared to be a basic principle of economy resembled the scientific thinking of the Enlightenment and some aspects of nineteenth-century science.

One should realize that Marx and Engels expected that the transition from capitalist to socialist society would be nonviolent and in accordance with democratic procedures. Because, according to Marxist theory, the changeover would take place in sociopolitically highly developed countries, the existence of the democratic infrastructure required for the nonviolent transition could be taken for granted. Only in countries lacking democratic institutions—for example, Russia—a radical, possibly violent change became necessary. Even the references to a "dictatorship of the proletariat" were not intended initially to suggest an authoritarian rule; Marx actually endorsed democratic forms of government at a time when the idea of democracy was by no means firmly established. According to Hook (1983, p. 21), Marx maintained that "any party, regardless of what it calls itself, that holds on to power by refusing to conduct free elections or to abide by their consequences is not Marxist." At this point, Marx's Communist Party, representing the labor movement as a whole, was still compatible with the sociopolitical complexity of democratic governance.

This conception of the Communist Party was easily perverted by Lenin during the Russian Revolution. Recall that after Russia's Czarist regime had toppled, a democratic form of government had been established in the first phase of that revolution, with a constitutional assembly in which the Mensheviks (headed by Alexander Kerensky) were the majority party; the communist Bolsheviks had received only nineteen percent of the vote. Lenin asserted that the Bolshevik Communist Party, as the only legitimate representative of Marxist communism, had to assume the leading role in the advance of the dictatorship of the proletariat. Counter to Marx's ideas, Lenin's dictatorship was based on "rule won and maintained by the use of violence by the proletariat against the bourgeoisie, a rule that is unrestricted by any law" (Hook 1983, p. 14). With this justification, Lenin forcibly dissolved the constitutional assembly, and claimed that the aims of the revolution were higher than the rights of the assembly. Subsequently, full centralization of governance and economy under the oppressive bureaucracy of the multinational Soviet Union completed this totalitarian simplification.

Communism is here suggested as a striking example of the inadequacy of sweeping simplifications imposed in place of complex reality. Even a superficial survey will suffice to reveal its shortcomings, many of which have been demonstrated time and again.

The assumption that sociopolitical reality can be fully defined by a minimal number of variables, comparable with the definition of the ideal systems of physics, is one of the most important fallacies of eighteenth- and nineteenth-century sociopolitical thinking. The establishment of these highly simplified forms of interpretation yields what could be called a "blinders-on-effect," a disregard of those factors invalidating the simplification.

One of the most fundamental oversights in a purely labor determined value of commodities is the role of how it is perceived psychologically. As a striking example, one can point to the price range of clothing, when the price of a one-of-a-kind designer dress may exceed the price of an analogous but mass-produced copy by one or several orders of magnitude, in spite of the similar amounts of work needed to produce two commodities. Capitalism has actually generated an entire profession—advertising—that may play havoc with Marxian labor-value theory, and thus artificially create a system of commodity values that depends to a large extent on psychological perception. The whims of imagination of a clever advertising expert may affect production in some sector of industry more effectively and be much more important than all of the alleged dialectical determinants. To carry the argument to its extreme, what would Marx have thought of the labor time-exchange value of the auction of a van Gogh painting to a Japanese bidder?

Contrary to Marxist prediction, the sociopolitically most advanced countries arrived at diverse, nonrevolutionary, non-Marxist solutions to the problematic relationship between labor and employer (see Chapter 5). The revolutionary Marxist-type takeover remained restricted to the sociopolitically lagging Russia and China and their satellites. Again, contrary to the simplifying dialectical prediction, the revolutionary takeover in Russia was carried out by a militant, highly organized minority, the Communist Party, not by a proletarian mass movement. The Russian and Chinese communist takeovers depended not only on the inherent forces of history, such as social conditions, but to a large extent depended on the initiative of Lenin's and Mao's unique and powerful personalities. Establishment of oppressive, privileged bureaucracies in communist Russia and China negated the prediction of a classless society as the final stage of Marxist historical dialectic. Also, Marx regarded the division of labor as one of the main shortcomings of the capitalist system. Although this form of production is a necessary result of industrialization, Marx did not suggest an alternative form of organization in which division of labor might be avoided or how productivity might be organized after the proletarian takeover and the establishment of a classless society.

The inadequacies and eventual failure of the communist system in Russia are emphasized here to suggest that abstract simplifications permit predictions only in the fully defined, ideal systems of physics. When dealing with nonideal systems—and sociopolitical systems that may be among the least ideal ones—the predictive value of abstract simplifications may be tentative and very limited. The state of these nonideal systems is determined by a large number of variables that cannot be accommodated in an abstract simplification. Instead, the aim must be to identify the largest possible number of factors affecting the state of a complex nonideal system, and to evaluate the relevance of individual factors under concrete and specific circumstances. Under Lenin, there was a tendency to consider a diversity of technical factors (such as rural electrification), as basic necessities for the maintenance of communist society; however, the introduction of a rigid centralization of the economy and collectivization of the peasantry as the main aspects of simplification generated a stagnation in economic development (excepting war- and space-related industry) that abolished the competitive position of Russia in the world economy and eventually led to its collapse.

The validity of this argument can be questioned in view of the persistence of communism in China and Cuba. Certain differences in the communist systems in Russia and China may be significant. To begin with, it should be remembered that Chinese communism has been in force for only about half as long as its Russian counterpart, and that any inadequacies have yet to show themselves. The question also arises: is it possible that in the Chinese system the nineteenth-century abstractions of Marx and Engels, as well as those in the Lenin-Stalin doctrines, were less significant as abstract simplifications than Mao's pragmatism? And, does the absence of collectivization of the Chinese peasantry indicate a less stringent elimination of individuality? Is this supported by the existence and recent promotion of private enterprise? Is the slow progress of the liberation of Chinese individuality the result of a more comprehensive evaluation of as many factors as possible in adjusting the precarious relationship of the individual to a society as vast and heterogeneous as China's?

An approach to these questions requires the exploration of complex nonideal model systems existing in nature or existing as societies with conspicuously extended continuity of social adjustment.

During the Enlightenment, simplifications of sociopolitical reality were conceived either as mechanistic representations of human existence or as social contracts. Both conceptions resulted from systems of thought without concrete practical implementation. In this sense, they were essentially utopias. In contrast, implementation of nationalism and materialistic dialectics followed rapidly upon their conceptualization, and took unanticipated forms and consequences. Thus, neither the nationalism that emerged from the French Revolution nor German nationalism developed through Fichte, Arndt, Jahn, and Hitler had been anticipated by Herder's initial conception of nationalism. Likewise, the communism of Lenin and Stalin went far beyond the Marx and Engels conception of dialectics. Also, nationalism and communism rejected conventional, mechanistic science as a basis for representing their sociopolitical realities. Nationalism did not substitute any rational scheme in support of its existence, and Marx and Engels introduced materialist dialectics as more adequately representing historical and physical reality than did conventional science, which they completely misunderstood.

Nationalism and communism are considered here as historical developments that simplified sociopolitical reality by modifying the delicate balance between individual and society, and between one society and the rest of humankind. Nationalism in its original Herderian sense envisaged a complex, pluralistic sociopolitical reality. By its uniqueness, each nation not only enhanced the individuality of its members, but it enriched humankind as a whole by heightening its diversity. A controlled nationalism actually intensified the experience of the individual as an independent whole and as a dependent part of the greater whole of a nation. Nationalism is still perhaps felt as the most deeply rooted extension of the self beyond its narrow boundary. Therefore, a foreigner within a nation is perceived not only as a threat to the unity of the nation but also to the integrity of the nationalistic individual, and thus results in xenophobia, and its violent forms of rejection.

The main danger of nationalism emerges when internal controls are lost. Apparently, at some point in its existence, every nation experiences an exaggeration of its self-esteem and an increase in its potential aggressiveness. As if to prove the validity of its nationhood, it sets out to conquer the rest of the world to replace the complexity of multinationality by a simplifying uni-nationality.

This inherent tendency toward expansionism is nationalism's most disastrous flaw. Not only the empire-builders of Rome, Spain, England, and France but even Denmark, Sweden, and Switzerland temporarily imposed the uniformity of a single nation in place of a more or less extended multinationality. The recent rampage of German nationalism, the horror of the Second World War, long in the making, was the most heinous case of this sociopolitical pathology.

In the Marxist representation of society, the individual loses temporarily his or her identity until the classless society comes into being. The existence of the bourgeois citizen is confined by greed, and the proletarian is alienated from his or her work by division of labor. During the revolutionary period the individual becomes submerged in the social superstructure. The absolute and utopian freedom of the individual in Marx's ultimate classless society (one of maximal complexity?) was actually inverted into an absolute suppression of individuality by an authoritarian bureaucracy (a case of maximal simplification?). Individual freedom and the integrity of the human environmental habitat were sacrificed to the drive for materialistic abundance for society as a whole.

The ultimate simplifications of nationalism and communism were enacted by Hitler and Stalin. They were said "to be driven by an inner force that was more than individual and represented a mysterious coincidence between personal egotism and mass will; they spoke both to the yearnings and to the fantasy of their age; they had the feel for power and the fine art of discriminating between its genuine forms and its counterfeits; they were utterly self-confident and walked unafraid above the abyss; and they were, above all, irreplaceable, transforming their worlds in ways that would have been inconceivable without them. . . . [They] had the gift of simplification that makes things seem transparent" (Craig 1992). Through Hitler's and Stalin's use of execution, imprisonment, and suppression of all civil rights, the complexities of individual human existence were obliterated more efficiently and more extensively than at any other time in history, and at the same time by usurping for themselves absolute and supreme individuality, both leaders became the most pernicious incarnations of ultimate historical simplification.

The liquidation of German destructiveness did not herald a worldwide end to nationalist aggression. Nationalistic demagoguery has flared up yet again in the Balkans and is rampant among the people of the Third World. The transition from socialist totalitarianism to economic and sociopolitical democracy is far from having reached its goal, and could well pass through phases of regression. It may be that the utopian abstractions of nationalism and socialist totalitarianism have withered away, but that their concrete implementation continues. Does the perception of a complex reality in democratic society that safeguards individual freedom offer an effective counterpoise to the totalitarian simplifications? At the close of the twentieth century, there is no certainty in affirmatively answering this question.

Chapter 7 Conclusion

The hope for relieving the burdensome and frustrating complexity of human existence through a simplified representation of reality, to be able to make complexity tolerable by subsuming it under a unifying system of thought, has been a pervasive goal of Western culture since the days of ancient Greece to the present. This aim has been nearly fully attained in the replacement of the diversity of physical phenomena by the theories that define the ideal systems of mechanics, thermodynamics, quantum mechanics, and relativity theory. We can speak about a full understanding of these systems because the theoretically related entities share general elements that provide a basis for the establishment of conceptual continuity. The main shared elements are force in mechanical systems; photons in the interaction of radiation and matter; molecular motility in the conversion of heat; and electrons in chemical reactions. This achievement is as essential for the understanding of reality as a grammar that provides the essentials for the use of a language is. But by themselves, the simplifying rules of the grammar do not tell us what words to use as subjects, verbs, and adjectives to produce the richness and specificity of a novel or a poem.

In the same sense, the general abstractions of the ideal systems of physics, the grammar of physical reality, do not tell what additional, specific variables have to be considered to understand how the nonideal systems of physics, biology, or sociopolitics persist and how their changes can be predicted with some probability. In terms of the abstract generalizations of physics, the planet earth is a mass of a certain magnitude that moves with a defined momentum around the sun. In terms of the concrete specificities of nonideal reality, the planet earth consists of a hard mantle, folded into mountains and beds of oceans and rivers covered by an atmosphere and a thin layer of living matter that includes the individual and collective human existence, all of them changing on multiple time scales.

At high levels of system-complexity the relevance of simplifying abstract generalizations declines because the number of variables is large and it is difficult, if not impossible, to identify those variables that in each moment alter the state of the system. This was clearly exemplified for biological systems by the review of muscle contraction and gene expression in which conceptual continuity is established by the identification of highly specific interactions between a great number of diverse system components.

Analogously, in the sociopolitical sphere, abstract ideologies and general theories are simplifications that do not lead to conceptual continuity. Ethnic, economic, or sociopolitical elements remain disconnected or even mutually destructive unless specific, possibly limited conditions can be generated, which are of mutual advantage and establish areas of conceptual continuity as a basis for coexistence. The question whether generalizations can be developed that are relevant for the understanding of the specific relationships of the elements of complex sociopolitical systems has not been conclusively answered.

Taking predictability as one of the main characteristics of ideal physical systems, a high degree of unpredictability in sociopolitical systems is one of the indicators of their complexity,[1] further emphasized by the specificity of multiple steps in the establishment of conceptual continuity between sociopolitical systems (concerning complexity of sociopolitical decision making, see Deutsch 1966; Voss, Lawrence, and Engle 1991).[2]

These developments suggest that humanity may have to abandon its dreams of simple realities and at last come to terms with the overriding complexity of both individual and social existence. The response to this change, the way we perceive reality, will very likely become an initial determining condition and characteristic of our future.

Three types of responses to complexity can be distinguished: resignation;

escape or denial; acceptance and attempted mastery. In the first case, due to a lack of understanding of the situation an individual passively succumbs to the complexity's impact and indiscriminately absorbs the concentrated excess of ubiquitous stimuli amplified by the mass media. Ultimately, this type of existence may give rise to various forms of dissatisfaction manifested as alienation (Murchland 1971), discontinuity (Drucker 1969) or uncertainty (Galbraith 1977).

Following the second option, escape or denial, an awareness of the complexity may motivate an escape from it. Thus, nationalism, fundamentalism, all of the various cults, intolerance, uncompromising rejection of cultural or ethnic diversity, and violence (including war) can be understood as simplifications of social complexities that offer ways of unloading an apparent burden of social responsibility (see Chapter 6).

Under the third option, acceptance and attempted mastery, complexity is accepted as a highly relevant form of sociopolitical reality, with democracy as its most pertinent example. A high level of complexity is created by a formal constitution, with its division of powers, checks and balances, its informally annexed political parties, its multiple advisory committees, and its implied promise of happiness for all citizens. The motivated and informed citizen is confronted with a wide range of choices and the need of making day-by-day decisions in reconciling the apparent disparities between competing demands and expectations from the private and the national and international public spheres of human existence. The challenge of these private-public opposites has become geometrically more compelling, and often disturbing, with the juxtaposition of competing approaches toward sociopolitical complexities, as presented in the ubiquitous media.

So far, the range of Western democracies—from the adversarial, competitive, majoritarian Americans, to the consociational Swiss, to the social democratic Scandanavians—have managed to ease the tensions between their private and public spheres. But these democracies have existed only for 200 years, and they are still based on the presumptions and conditions that prevailed at the time of their origin. Democracies now face changes that seem more radical than those encountered at any preceding period, namely, the structural unemployment, the requirement for economic growth with concomitant increase in consumption, pollution, and waste, all of which are incompatible with our limited resources and the maintenance of a livable environment. In addition, there is the sociopolitical takeover by powerful, national, and global corporations that threatens the basic structure-function of democracy (Soros 1997).

In the last half of the twentieth century, it has become problematical as to whether or not democracy's high level of complexity can be successfully maintained over an extended historical period. This question has been answered negatively from an archeological-historical perspective (Butzer 1980; Tainter 1988); from a sociopolitical perspective (Crozier, Huntington, and Watanuki 1975; Attali 1990; Zolo 1992; Bethke-Elshtain 1995); from a socioeconomic perspective (Kennedy 1987, 1993) and from a perspective of the global situation as whole (Kaplan 1994; Huntington, 1996). In particular, the precariousness of the American adversarial-competitive-majoritarian democracy has been emphasized (Kennedy 1993). But, the perseverance of democratic complexity is affirmed by Chancellor 1990; Nye 1991; Nau 1992 (from Kennedy, 1990); and Ladd, 1993a, 1994. It can be asked: does the complexity of American democracy hold promise for adjustment to the exigencies of the present time, or is it particularly vulnerable to factional disputes and partisan adversarial squabbles that delay or prevent the making of decisions? In this way the point of no return may be missed for taking those steps that are necessary for the long-range maintenance of human society without violence and without destruction of the environment as a habitat. Are there any special resources available to counteract such deleterious trends?

On several occasions, American democracy met serious challenges by shifting party dominance over the public/private oppositeness. The critical elections of 1800, 1828, 1860, 1896, 1932, 1964, and 1980 each had such a result. Can problems of the magnitude facing democracies today still be resolved by the dominance of alternating parties? Since the Western democracies presently still are playing a leading role in the world, the question is whether the adversarial-competitive-majoritarian form or the consociational form of the democratic spectrum will be more effective in coping with our grievous future problems, or will the future proceed in the thrall of quasi-democractic or authoritarian forms of governance? Is it then legitimate to emphasize known variables or add new ones to its conceptual inventory?

As a reply, this book has not been written with the pretense of finding actual solutions to these problems. It merely calls attention to the dangers of using simplifications of reality as models of human sociopolitical existence. This applies to both the use of the abstract generalizations, such as Newtonian mechanics for interpreting human society, and to concrete ideological implementations of abstract simplifying sociological utopias, such as fascism and communism. It is commonly the citizen's preferred option to seek simple solutions of complex sociopolitical problems and to adopt one side of a sociopolitical dichotomy instead of recognizing and adjusting to a conceptual continuity between both sides of a duality. It is a matter of primary importance to prepare at least an articulate minority of citizens to accept complexity as the relevant mode of sociopolitical reality and to replace the perception of mutually exclusive components of dichotomies by the inclusiveness of duality. Is it possible to reinforce the duality of the individual—the quality of being at the same time a distinctive social entity and a part of the greater whole of society and of mankind? Should some form of social service in the community or in the Peace Corps become more obligatory to enhance the perception of the relatedness of the private/public components of sociopolitical reality?

As a promising answer to these problems, deliberative participatory democracy has found much attention. It was heralded by Barber (1984) and broadly elaborated by Habermas (1996) and Gutmann and Thompson (1996). This form of democracy is here set apart from the alternatives of liberalism and communitarianism introduced in Chapter 5. In contrast to the abstract, philosophical-academic origin of the latter forms, deliberative democracy has deep roots in American history in the form of the town meetings of the early nineteenth century, which impressed de Toqueville as an outstanding characteristic of American society in those days. In its contemporary development "the gap between the theory and practice of deliberative democracy is narrower than in most other conceptions of democracy" (Gutmann and Thompson 1996, p. 357). According to these authors, the principles of reciprocity, publicity, and accountability regulate deliberation with liberty, basic opportunity, and fair opportunity as main components of deliberative subject matter.

Accordingly, deliberative democracy is not expected to decide the validity of abstract sociopolitical categories. Deliberative democracy attempts to directly establish some understanding between opposing civic groups, to find areas of common interest, and to generate joined projects. It aims at actively involving the citizen at large in facing sociopolitical problems, in trying to come to an understanding of the major issues of the time, and in finding compromise solutions. In this sense, deliberative democracy leads to the implementation of conceptual continuity. Actually, restating the basic tenets of deliberative democracy and refuting its critics, Bohman advocates the persistent and determined improvement of the methods and techniques of public discussion as a basic requirement for the maintenance of democracy (1996). It is noteworthy, therefore, that the realization of this requirement is proving highly successful. A far-flung network of "study circles"[3] as been created that provides opportunities for the common citizen to replace the adversarial aspects of the issues of race relations, health care, welfare, the role of the United States in world conflicts, by the recognition of common interests between the opposing parties. In the course of the study circle discussions, the balancing of the private and public spheres of existence engenders the experience of the duality of the individual that requires a heightening of the mental effort. As reward, the awareness of this duality, of our lives as individual selves and as interrelated parts of society, generates a double-life experience in a single-life time. Without this adjustment to complexity virulent nationalism—or at least the Singapore model of a benevolent dictatorship—looms on the horizon as the remaining alternative. The message preferably to be followed into the twenty-first century has been proclaimed for some time by I. Berlin (1991):

> Reality is not predicated upon mutually exclusive alternatives but upon the inclusive belief that individuals can understand, even sympathize with the disparate beliefs and values of others. Its result is not polarization but pluralism, the idea that there are many different ends that men may seek and still be fully rational . . . capable of understanding each other.

The establishment of conceptual continuity is here introduced as a prerequisite of understanding complex pluralistic systems, even the multiple manifestations of human existence.

Notes

1. The recognition of a simple/complex dichotomy in the perception of reality has become apparent in different contexts: Pondering the quantum mechanically defined actions of photons, the elementary energy units of radiation, Feynman comments:

"You might wonder how such simple actions could produce such a complex world. It's because phenomena we see in the world are the result of an enormous intertwining of tremendous numbers of photon exchanges and interferences. Knowing the three fundamental actions is only a very small beginning toward analyzing any real situation where there is such a multitude of photon exchanges going on that it is impossible to calculate—experience has to be gained as to which possibilities are more important. Thus we invent such ideas as 'index of refraction' or 'compressibility' or 'valences' to help us calculate in an approximate way when there is an enormous amount of detail going on underneath. It's analogous to knowing the rules of chess which are fundamental and simple—compared to being able to play chess well, which involves understanding the character of each position and the nature of various situations—which is much more advanced and difficult" (Feynman 1988, p. 114).

Reviewing Toynbee's monumental generalization of world history, G. F. Kennan concludes that "the life of every society is marked by thousands of

211

circumstances; and to attempt to press this multitude of variables into the Procrustean bed of a great general system embracing all societies seems to me an unnatural and unpromising undertaking." Kennan is wondering whether it was "Toynbee's failure to take full account of the element of the fortuitous—of pure chance—in the unfolding of human affairs, and his consequent assumption that everything that happened in those affairs had to have a reason visible and intelligible to the human eye" (Feynman 1988, p. 20) and that, as Kennan continues, "at the bottom of all human experience there lay, after all, the mystery of the individual personality—its ultimate autonomy of decision—its interaction with the mass" (Kennan, *New York Review of Books,* June 1, 1989, p. 21). A similar point of view has been set forth by I. Berlin (1996). Finally, in his *Grammatical Man,* J. Campbell (1982) emphasizes the shift in relevance from the general to the specific by pointing to the importance of the general rules of grammar as an essential basis of language to their relative unimportance in our perception of the literary and aesthetic quality of a novel or a poem. General laws that apparently control chance events in nature and in games have been examined in great depth by Eigen and Winkler (1981).

2. Several attempts have been made to give complexity—or at least some aspects of it—a formalized general interpretation, such as systems theory, information theory, chaos theory, and theories of dissipative irreversible thermodynamics. It remains uncertain, however, whether singly or together these theories lead to conceptual continuity (for references and further discussion see Chapter 4). As discussed in more detail in Chapter 4, it is not suggested that the general abstractions of fully defined, ideal systems are irrelevant to the representation of partially defined, nonideal systems. Instead it means that in the ideal systems of physics general theories lead directly to conceptual continuity. In contrast in systems that consist of a large number of heterogeneous components with highly specific interactions such general theories, including theories of complexity, do not lead to conceptual continuity. In such systems conceptual continuity depends on the identification of these highly specific interactions. This is exactly what has happened in cell and molecular biology and is also happening in sociopolitical systems.

3. At the same time it is important to realize that some of the most fundamental mathematical formalizations of physical reality—Newtonian mechanics as well as certain basic electric phenomena—do not yet represent full conceptual continuity. In Newtonian mechanics gravitational attraction is the system component that links two masses and is the basis for the formal relationship between two masses. It is not fully understood, however, how gravitational force produces attraction between two masses and therefore conceptual continuity is not fully established (this shortcoming of the Newtonian system was anticipated by Locke, see pages 90–91). Similarly, even in the routine observation of attractions and repulsions of bodies with opposite or identical electric charges respectively, the nature of the interaction of the electric fields between the two bodies and the atoms of the interacting bodies is not fully understood. Therefore, conceptual continuity is still wanting, in spite of the elementary nature of this phenomenon (see also Chapters 2 and 3).

4. A recent discussion of the origin of Western science is provided in a review by J. North in "Why Western Europe: The Danger of Seeking a Single Explanation for the Rise of Science" *Times of London Literary Supplement* XII, December 15, 1995, p. 3).

5. The full scope and relevance of the problem of complexity and the diversity of approaches

for its elucidation are indicated by the following quotations from a Symposium on "The Science and Praxis of Complexity" (Aida et al. 1985).

"Complexity cannot be managed, intellectually or practically, through increased control. We will have to learn to understand and manage complex systems while respecting the autonomy of the processes and the elements within these systems. We will have to rethink our attitudes towards order and disorder, and accept that disorder is not only negative but also a precondition for the creation of new orders. We need to consider the levels of disorder that we can accommodate in a human manner without recourse to oppression and violence. And for this task we need to draw upon the insights of all cultures" (Soedjatmoko 1985, pp. 5–6).

"The problem of complexity is not one of completeness, but rather of incompleteness of knowledge. In a sense, complex thought tries to take account of what is discarded and excluded in the mutilating type of thoughts that I call simplifiers, and thus it combats not incompleteness but mutilation. For example, if we think of the fact that we are physical, biological, social, cultural, psychic and spiritual beings, complexity is obviously that which attempts to link or identify these aspects by highlighting the differences between them, whereas simplified thought either separates these different aspects or unifies them through a mutilating reduction" (Morin 1985, p. 63).

CHAPTER 1: THE DAWN OF THE SIMPLE-COMPLEX DICHOTOMY AND OF CONCEPTUAL CONTINUITY

1. In the sociopolitical context the idea of Conceptual Continuity is related to the consociational relationship as suggested in Chapter 5.
2. The characteristics of disequilibrium again acquires prominence in the discussion of complex physical, and in particular living systems.
3. That Athens should not be regarded as an out-and-out humanitarian community has been poignantly set forth by Sagan in the opening pages of his *Honey and Hemlock* (1991, p. 2).

   "On the one hand, quite aside from tragedy, philosophy, psychology, the study of politics and science, Athens (not our other ancient ancestor, Rome) has given us the greatest gift imaginable: the ideal and the reality of a democratic policy based on a complex and moral conception of citizenship. On the other hand, with the hand that held the sword, Athens bequeathed a cruel and imperial domination of other Greek cities, the slaughter and enslavement of its wartime opponents, the occasional genocide of another polis, not to mention the ownership of tens of thousands of domestic and industrial slaves and the almost total exclusion of women from cultural and political life. It is of crucial importance to try to understand what gross immoralities are still compatible with the forms of democratic society. Athens provides us with one of the sharpest examples, if not the sharpest, of this awesome human contradiction."

4. Plutarch wrote that "Lycurgus trained the citizens neither to wish nor to be able to live as individuals. Like bees, they were always to be integrated with the state swarming round their leader, almost beside themselves in their eagerness and rivalry to belong wholly to the state" (Plutarch, Lycurgus, quoted by Moore. Aristotle and Xenophon on Democracy and

Oligarchy. Translations with Introductions and Commentary, University of California Press, Berkeley, 1975, p. 103).

"People of antidemocratic, authoritarian temperament have been attracted to, and approving of, the Lacedemonian state. Even during the Peloponnesian War some Athenian youths of oligarchic inclinations wore their hair long in imitation of the Spartan style. Critias, we have observed, was a great enthusiast of the Spartan constitution. The list of Sparta's admirers includes Xenophon, Plato, and several twentieth century enthusiasts of the totalitarian state" (Sagan 1991, p. 149).

5. The limits of this tolerance were obviously too narrow for the nonconformity of Socrates.

6. In the second book of *Politics,* Aristotle attempts to find a balance between the need for social stability and the need for social change. On the one hand, maintenance of tradition is desirable because it is the result of trial and error over long periods in the history of societies. On the other hand, the possibility for change is essential to adapting laws to new circumstances, for example, the need for providing adequate supplies of food for the poor in years of inadequate harvests. Furthermore, it is possible to compare the merits of the traditions of different societies, and Aristotle attempted to reconstruct the optimal social organization for the achievement of the good life (eudaimonia) for each individual. What he calls virtues, such as courage, can be regarded as stable traits of all societies (see also M. Nussbaum, *New York Review,* December 7, 1989, *Recoiling from Reason*).

CHAPTER 2: THE REASSERTION OF THE SIMPLE-COMPLEX DICHOTOMY IN THE MODERN ERA

1. Theology gave rise to an important form of the concept of "Natural Law" as indicated in the following quotation:

"Since all things subject to Divine providence are ruled and measured by the eternal law . . . it is evident that all things partake somewhat of the eternal law, insofar as, namely, from its beginning imprinted on them, they derive their respective inclinations to their proper acts and ends. Now among all others, the rational creature is subject to Divine providence in the most excellent way, insofar as it partakes of a share of providence, by being provident both for itself and for others. Wherefore, it has a share of the eternal reason, whereby it has a natural inclination to its proper act and end; and this participation of the eternal law in the rational creature is called the natural law" (St. Thomas, *Summa Theologica,* Part II [First Part, Q.XCI, Art. II] From C. L. Becker. *The Heavenly City of the Eighteenth-Century Philosophers,* Yale University Press, 1932, p. 3).

In Becker's words:

"Nature seemed inaccessible, mysterious, and dangerous, at best inharmonious to men. Men therefore desired some authoritative assurance that there was no need to be apprehensive; and that this assurance came from theologians and philosophers who argued that since God is goodness and reason, his creation must somehow be, even if not evidently, so to finite minds, good and reasonable. Design in nature was thus derived a priori from the character which the Creator was assumed to have; and natural law, so far from being associated with observed behavior of physical phenomena, was no more than a conceptual universe above and outside the real one, a logical construction dwelling in the mind of God and dimly reflected in the minds of philosophers" (Becker p. 55).

2. "Look around the world: contemplate the whole and every part of it: You will find it to be nothing but one great machine, subdivided into an infinite number of lesser machines, which again admit of subdivisions, to a degree beyond what human senses and faculties can trace and explain. All these various machines, and even their most minute parts, are adjusted to each other with an accuracy, which ravishes into admiration all men, who have ever contemplated them. The curious adapting of means to ends, throughout all nature, resembles exactly, though it much exceeds, the productions of human . . . intelligence. Since therefore the effects resemble each other, we are led to infer . . . that the causes also resemble; and that the Author of Nature is somewhat similar to the mind of man; though possessed of much larger faculties, proportioned to the grandeur of the work, which he has executed" (Hume, *Dialogues Concerning Natural Religion,* 1907, p. 30; quoted in Becker 1933, p. 56).

3. In the present context the idea of conceptual continuity is introduced to give the term mechanism a more precise meaning. It is apparent from the foregoing quotation that the mathematically formalized, general relationships of ideal physical systems postulate general forms of conceptual continuity.

4. In a further elaboration T. M. Porter maintains: "Since the time of Galileo and Bacon, and especially since Newton, natural philosophy has stood for the possibility of attaining reliable, objective knowledge. As a model for social thought, it offered a possibility of escape from the tyranny of faction and partisan politics. Scientists, it appeared, are somehow able to achieve consensus about the fundamentals of the issues they study. For more than three centuries natural science has served as a prototype for the authority of reason which seemed an agreeable contrast to the regime of status, wealth and prejudice" (Porter 1990, p. 1024).

   But as a caveat Porter adds: "It is especially significant that the natural sciences present neither a unified nor any single readily applicable model for social science. Science envy, far from elevating political and social thought above politics has provided instead a pervasive idiom of debate" (Porter 1990, p. 1024).

5. Short biographical annotations may provide backgrounds for the intellectual developments of the personalities followed in this chapter.

6. Hobbes spent the years from 1636–37 in Padua, where Galileo and Harvey were then working.

7. Herbert (1989, p. 5) points out: "His [Hobbes's] own clearly stated intention was to produce a single coherent, all-inclusive system of philosophy—to be called the Elements of Philosophy—grounded on the principles of natural science, progressing systematically through the science of human nature to its culmination in a science of the principles of civil association. The three elements which together were to constitute his system were defined in his 'De Corpore'; 'De Homine' (De Natura) and 'De Cive.'"

   Although Herbert questioned the coherence of Hobbes's system by pointing to possible conceptual discontinuities between his mechanistic interpretation of nature and man's psychology, or between the latter and the conception of governance, Herbert upheld the necessity for an integrating representation of Hobbes's work.

"The principal problem was the major reinterpretations of Hobbes's philosophy in that they ignore the very problem to which Hobbes's philosophy expressly addresses itself, that is, the problem of providing a theory of man and the world which does not divide the world into a variety of unrelated parts, which does not leave unexplained any major part of appealing to some principle (say, 'separated essences') that is outside the theory and consequently outside the world as well. . . .

"A genuine interpretation of Hobbes's thought must attempt to integrate the parts of his philosophical system in a manner consistent with this intention" (Herbert 1989).

8. The metaphorical nature of Hobbes's science is indicated by the opening of his main work, *Leviathan,* that summarizes and relates his separate lines of thought.

The Introduction to Leviathan *by Hobbes*

NATURE (the Art whereby God hath made and governes the World) is by the *Art* of man, as in many other things, so in this also imitated, that it can make an Artificial Animal. For seeing life is but a motion of Limbs, the begining whereof is in some principall part within; why may we not say, that all *Automata* (Engines that move themselves by springs and wheeles as doth a watch) have an artificiall life? For what is the *Heart,* but a *Spring;* and the *Nerves,* but so many *Strings;* and the *Joynts,* but so many *Wheeles,* giving motion to the whole Body, such as was intended by the Artificer? *Art* goes yet further, imitating that Rationall and most excellent worke of Nature, *Man.* For by Art is created that great LEVIATHAN called a COMMON-WEALTH, or STATE, (in latine CIVITAS) which is but an Artificiall Man; though of greater stature and strength than the Naturall, for whose protection and defence it was intended; and in which, the *Soveraignty* is an Artificiall *Soul,* as giving life and motion to the whole body; The *Magistrates,* and other *Officers* of Judicature and Execution, artificiall *Joynts; Reward* and *Punishment* (by which fastned to the seate of the Soveraignty, every joynt and member is moved to performe his duty) are the *Nerves,* that do the same in the Body Naturall; The *Wealth* and *Riches* of all the particular members, are the *Strength; Salus Populi* (the *peoples safety*) its *Businesse; Counsellors,* by whom all things needfull for it to know, are suggested unto it, are the *Memory; Equity* and *Lawes,* an artificiall *Reason* and *Will; Concord, Health; Sedition, Sicknesse;* and *Civill war, Death.* Lastly, the *Pacts* and *Covenants,* by which the parts of this Body Politique were at first made, set together, and united, resemble that *Fiat,* or the *Let us make man,* pronounced by God in the Creation.

To describe the Nature of this Artificiall man, I will consider

First, the *Matter* thereof, and the *Artificer;* both which is *Man.*

Secondly, *How,* and by what *Covenants* it is made; what are the *Rights* and just *Power* or *Authority* of a *Soveraigne;* and what it is that *preserveth* and *dissolveth* it.

Thirdly, what is a *Christian Common-wealth.*

Lastly, what is the *Kingdome of Darkness* (Hobbes 1973, p. 1).

9. For Hobbes, "science" is the knowledge of consequences (Hobbes 1975, p. 21):

"By this it appears that reason is not, as sense and memory, born with us; nor gotten by experience only, as prudence is; but attained by industry; first in apt imposing of names; and secondly by getting a good and orderly method in proceeding from the elements, which are names, to assertions made by connection of one of them to another; and so to syllogisms, which are the connections of one assertion to another, till we come to a knowledge of all the consequences of names appertaining to the subject in hand; and that is it men call 'science'. And whereas sense and memory are but knowledge of fact, which is a thing past and irrevocable, 'Science' is the knowledge of consequences, and dependence of one fact upon another; by which, out of that we can presently do, we know how to do something else when we will, or the like another time; because when we see how anything comes about, upon what causes, and by what manner; when the like causes come into our power, we see how to make it produce the like effects."

However, Hobbes expresses the following reservation: "As for the knowledge of consequences, which . . . is called a science, it is not absolute, but conditional. No man can know by discourse, that this, or that, is, has been, or will be; which is to know absolutely; but only, that if this be, that is . . . ; which is to know conditionally" (Hobbes 1975, p. 42).

10. Herbert identifies a central issue: "In his most mature natural philosophy, Hobbes identified motion, or more specifically, the most fundamental and minute constituents of natural motion, with conatus. The result is a dynamic and dialectical conception of nature according to which all difference and all individuation is the product of dissociating forces, which resist and even oppose one another, individuating themselves in the process.

"Conatus is an extension of that dynamic, dialectical impulse that is the rudiment of all human thought and action, and, in fact, of all nature" (Herbert 1989, p. 64).

11. Of particular importance in the entire system of Hobbes's thought is that the concept of motion stands for any form of change. Motion is the result of mechanical interactions that are mediated either directly or through some medium as for example, in the transmission of light (Kraynak 1990, p. 129). Hobbes's interpretation of light is instructive (Herbert 1989, p. 38, quoting Hobbes):

Hobbes's Scheme of Consequences. "So that light is nothing but a fancie (phenomenal object; Herbert 38) made by the lucid object by such pressure as I have even now described but this pressure is really and actually a local motion of the parts, both of the lucid object which comes a little forward every way, and also of the organ, that is to say, of the spirits of the heart, and yet parts of the brain and of the optique nerve (though the said motions be imperceptible) and is not a mere inclination. For all inclination if it be pressure or endeavour is actually a motion and progression of something out of its place, pressure cannot otherwise be conceived."

Herbert comments on that passage:

"Hobbes appears to be arguing that bodies whether luminous or not, are distinguished from others, that is, are determinate, because of internal motions, and that light and vision occur because of the reciprocal determinacy of motions. Later in De Corpore, Hobbes will define light as an 'endeavour outward' caused by the reaction of the heart to pressures propagated from outside (De Corpore, Ch. 27, art. 2, p. 448). This makes light a fancie, that is a phenomenal object.

"It is the interaction between a luminous object and the recipient that makes the luminous body determinate (determinate = having defined limits)" (Herbert 1989).

12. "Therefore it is necessary in entering into such a covenant for its members . . . to conferre all their power and strength upon one Man, or upon one Assembly of men, that

TABLE N.1 Hobbes's System of Consequences

SCIENCE, that is, Knowledge of Consequences; which is called also PHILOSOPHY.

- **Consequences from the Accidents of Bodies Naturall; which is called NATURALL PHILOSOPHY.**
  - **Consequences from all Accidents common to all Bodies Naturall; which are Quantity, and Motion.**
    - Consequences from Quantity, and Motion *indeterminate*; which being the Principles, or first foundations of Philosophy, is called *Philosophia Prima*. → **PHILOSOPHIA PRIMA.**
    - Consequences from Motion, and Quantity *determined.*
      - *Mathematiques,*
        - By Figure, ... → **GEOMETRY.**
        - By Number, ... → **ARITHMETIQUE.**
      - Consequences from the Motion, and Quantity of Bodies in *speciall.*
        - Consequences from the Motion, and Quantity of the great parts of the World, as the *Earth* and *Starres.* — *Cosmography,*
          - → **ASTRONOMY.**
          - → **GEOGRAPHY.**
        - Consequences for the Motion of special kinds, and Figures of Body, — *Mechaniques, Doctrine of Weight,*
          - → *Science of* **ENGINEERS.**
          - → **ARCHITECTURE.**
          - → **NAVIGATION.**
          - → **METEOROLOGY.**
  - **PHYSIQUES, or Consequences from Qualities.**
    - Consequences from the Qualities of Bodies *Permanent.*
      - Consequences form the Qualities of Bodyes *Transient,* such as sometimes appear, sometimes vanish
        - Consequences from the Qualities of the *Starres,*
          - Consequences from the *Light* of the Starres. Out of this, and the Motion of the *Sunne,* is made the Science of . → **SCIOGRAPHY.**
          - Consequences from the *Influence* of the Starres → **ASTROLOGY.**
        - Consequences of the Qualities from *Liquid* Bodies that fill the space between the Starres; such as are the *Ayre,* or substance aetheriall.
      - Consequences from the Qualities of Bodies *Terrestriall.*
        - Consequences from the parts of the Earth, that are without *Sense,*
          - Consequences form the Qualities of *Minerals,* as *Stones, Metalls,* &c.
          - Consequences from the Qualities of *Vegetables.*
        - Consequences from the Qualities of *Animals.*
          - Consequences from the Qualities of *Animals in generall.*
            - Consequences from *Vision,* → **OPTIQUES.**
            - Consequences from *Sounds,* → **MUSIQUE.**
            - Consequences from the rest of the *Senses.*
          - Consequences from the Qualities of *Men in speciall.*
            - Consequences from the *Passions of Men* → **ETHIQUES.**
            - Consequences from *Speech,*
              - In *Magnifying, Vilifying,* &c. → **POETRY.**
              - In *Persuading,* → **RHETHORIQUE.**
              - In *Reasoning,* → **LOGIQUE.**
              - In *Contracting.* → The *Science of* **JUST and UNJUST.**
- **Consequences from the Accidents of *Politique* Bodies; which is called POLITIQUES, and CIVILL PHILOSOPHY.**
  - 1. Of Consequences from the *Institution* of COMMON-WEALTHS, to the *Rights,* and *Duties* of the *Body Politique,* or *Soveraign.*
  - 2. Of Consequences from the same, to the *Duty,* and *Right* of the *Subjects.*

Source: Thomas Hobbes, *Leviathan,* 1973, p. 40. Courtesy Everyman's Library.

may reduce all their Wills by plurality of voices, unto one Will; which is as much as to say, to appoint one Man, or Assembly of men, to bear their Person; and every one to owne, and acknowledge him their Person, shall Act, or cause to be Acted, in those things which concerne the Common Peace and Safetie; and therein to submit their Wills, every one to his Will, and their Judgment, to his Judgment. This is more than Consent, or Concord; it is a reall Unitie of them all, in one and the same Person, made by Covenant of every man with every man, in such a manner, as if every man should say to every man. I Authorise and give up my Right of Governing my selfe, to this Man, or to this Assembly of men, on this condition, that you give up thy Right to him, and Authorise all his Actions in like manner. This done, the Multitude so united in one Person is called a Common-Wealth, in latin Civitas. This is the Generation of the great Leviathan, or rather (to speak more reverently) of that Mortal God, to which we owe under the immortal God our peace and defence" (Hobbes 1975, *Leviathan*, p. 89). "And he that carryeth this Person, is called Soveraigne, and said to have Sovereign Power" (p. 90).

13. See note 10 above.

14. "Recognition of the double nature of social man who is both an individual and a citizen, the persistence of natural law in the contractual state" (Leigh 1988, p. 243).

"Here, as at so many other occasions, Rousseau stands with one foot in each camp. He never outgrew the methods of logical, intellectual analysis learned from the Philosophes. But he differs from them in that they created man in their own image by combining Cartesian logic with Lockian utilitarianism. On the whole, despite exceptions that can easily be pointed out, it is true to say that they allowed only for the intellectual, the conscious and utilitarian motives in human psychology. They are individualists, but their individuals are all cast in the same mold; they proclaim man's freedom, but only insofar as he behaves as they think he should. Rousseau's is a different brand of individualism. His 'return to nature' upset all this, for by nature he meant the nature of the individual man, in all its variety, idealized by being moralized and sentimentalized, but not transformed, into a homogeneous and uninteresting mass by being strained through the fine mesh of encyclopedic intellect. Where the Philosophes adored uniformity, Rousseau preferred all the individual, local, and national variations that so complicate problems for the students of society" (Cobban 1934, pp. 236–237).

"The Philosophes pushed Locke's ideas to conclusions which he himself would hardly have dreamed of, but they introduced no new principles, and in practice most of them gave evidence of extreme timidity. The historian has sounded the death-knell of many a good revolutionary reputation and even Robespierre himself is now held up to scorn as the embodiment of petty bourgeois prejudices.

"In Rousseau the combination of the revolutionary and the conservative is unusually clearly marked, and each element contributed its quota to his political thinking. But since revolution implies change and conservatism repose, his revolutionary principles were a call to action, his conservatism an aid to understanding. There is no conflict between the two sides of his mind. On the contrary, they are intimately connected, for without the insight into the deep-lying forces of human nature which is the source of his conservative tendencies, what was really original, what indeed was most truly revolutionary in his political philosophy could never have existed and Rousseau would have counted as one of the rank and file Philosophes and no more" (Cobban 1934, pp. 238–239).

Rousseau presented us with two utopias: "One utopia rested on an ideal of individual autonomy, and it stresses the need for a countercultural education, private family life, and countermodern institutions to protect the individual from the modern worlds and empires of opinion, error, oppression, inequality, and greed. This private educative utopia is spelled out in *La Nouvelle Heloise* and *Emile,* the seminal statements of the private, countercultural and therapeutic ideal. The other utopia is civic, political, and collective. It is found in works like the *Social Contract,* and its underlying image is that of the city state, where the individual finds unity by merging himself with the civic unit-later that nation, the party, the movement" (Featherstone 1978, pp. 186–187).

15. The main condition maintaining the natural state was already pointed out by Durkheim more than a century ago: "The perfect balance between the needs of man in the natural state and the available resources. . . . Man's desires do not go beyond his physical needs. . . . His needs are easily satisfied. Nature has provided for them. . . . Man is in harmony with his environment because he is a purely physical being, dependent on his physical environment and nothing else. The nature within him necessarily corresponds to the nature without" (Durkheim 1960, pp. 70, 71).

16. Perhaps realizing the difficulty in conveying the meaning of the concept of "General Will," Rousseau attempts a clarification in his discourse *On Political Economy.* "Permit me for a moment to use a common comparison, inaccurate in many respects but suited to making myself better understood.

"The body politic, taken individually, can be considered as an organized, living body and similar to that of a man. The sovereign power represents the head; the laws and the customs are the brain, the center of the nervous system the seat of understanding, the will, and the senses, of which the judges and the magistrates are the organs; commerce, industry, and agriculture are the mouth and stomach which prepare the common subsistence; public finances are the blood that a wise economy, performing the functions of the heart, sends back to distribute nourishment and life throughout the body; the citizens are the body and members which make the machine move, live and work, and which cannot be injured in any way without a painful sensation being transmitted right to the brain, if the animal is in a state of good health.

"The life of both (meaning sovereign and the citizen?) together is the *self* common to the whole, the reciprocal sensibility and the internal connection between all the parts. What if this communication ceases, the formal unity disappears, and the contiguous parts are only related to one another by their juxtaposition? The man is dead, or the state is dissolved.

"The body politic is, therefore, also a moral being which has a will, and this general will, which always tends toward the conservation and welfare of the whole and of each part, and which is the source of the law, is, for all the members of the state, in their relations to one another and to the state, the rule of what is just and unjust, a truth" (Ritter and Bondanella 1988, p. 61).

This general will is not an absolutist concept and may differ from society to society: "It is important to note that this rule of justice, which is unerring with respect to all citizens, may be faulty with respect to foreigners, and the reason for this is obvious. The will of the state, although general in relation to its members, is no longer in relation to other states and their members, but becomes for them a particular and individual will, which has its rule of justice in the law of nature (notice use of Law of Nature). This enters equally into the principle already established, for the great city of the world then becomes the body politic whose law of nature, as always the general will, and whose states and various peoples are merely individual members" (Ritter and Bondanella 1988, p. 62).

17. Explicit claims by eighteenth- and nineteenth-century critics, and more recently by Talmon (1952), of Rousseau's role as one of the originators of totalitarianism were rejected by Leigh (1978, pp. 242–243).

18. Rousseau was not a dogmatic critic of the English parliamentary system. He merely objected to certain specific shortcomings in the relationship between the representatives and their constituents. He would have accepted representation were the representatives compelled to follow more closely the mandates of the constituency (Cobban 1934, pp. 62–63).

The issue of representation is one of the main points of disagreement between Rousseau and Burke. The latter regarded representation as a hereditary privilege of the landed aristocracy whereas Rousseau insisted on representation through election by an unrestricted constituency (Cobban 1934, p. 64).

19. Rousseau's associations can perhaps be compared to our contemporary political action committees or to President Eisenhower's military industrial complex (after Cobban 1934, p. 70).

20. And in an alternative rendition: "Rousseau's eupsychic legacy is the fantasy of a perfectly autonomous, fulfilled 'I' for every man, the wholeness of a communal 'I' that is an organic unity, and the integration of the entire, individual 'I' with the communal 'I' with hardly a ripple on the surface" (Manuel 1970, p. 2).

Rousseau himself saw the complexity implied in his utopian dreams:

"Richly endowed with a utopian propensity, Rousseau confronted some of the prickly questions that characterize the Western utopian way in both argumentative discourse and description. The degree [notice quantitation] of symphasis, or cohesion, that should exist among individuals in society is probably the most deep-rooted utopian problem in the rationalist tradition; in what kind of social unit will this ideal find appropriate expression? What is the place of the ideal world as an age in the eternal passage of time? Granted the crucial significance of sexuality in mankind, what is the optimal pattern of relationship between the two parts of the species? Since education is the key to any society, what is the perfect model for upbringing? What should be the relationship between need and desire? What happens to individuality in the communal utopia? Is there an ideal religious spirit that should be infused into the feeling tone of society? What part of man's nature should by preference be perfected, his moral or his intellectual faculties?" (Manuel 1978, p. 2)

And further on: "Rousseau's utopian 'I' has been embraced by a branch of psychology and has been spread about in multiple versions; the communal 'I' is a political dogma taught over half the globe and their fusion is part of a world revolutionary credo" (Manuel 1978, p. 2).

Rousseau's utopia of the individual "I" indicates one of the reasons for his rejection of science again as suggested by Manuel (p. 5):

"The ideal '*moi*' has harmoniously educated manual and mental powers; it cannot conceive of luxuria; it is autonomous, entire, whole. It lives fully and totally within the bounds of time and space that happen to be its environment. Identity, the consciousness of self, grows like a plant. Once it is fashioned, man can preserve that self whatever the vicissitudes of fortune. The limitless development of the ideal '*moi*' in scientific knowledge and the quest for unending novelty only deprive the '*moi*' of the immediacy of joy in the present. Rousseau had a grave suspicion of the pleasures of dominion, either over persons or over nature. The intellectual pleasure of scientific curiosity was forthrightly rejected—Rousseau did not even bother to argue the point seriously. It was Auguste Comte in his second career as High Priest of humanity who took a similar position a century later and gave 'physiological' reasons for the superiority of an expansive, sentient *moi* over the rationalist *moi* dedicated to scientific inquiry" (Manuel 1978, p. 5).

21. The wide range of interpretations dealing with Rousseau's political writings is indicated by T. B. Strongs (*American Political Science Review,* 1996, 90: No. 4) in a review of two books: J. Simon, 1995, *Mass Enlightenment: Critical Studies in Rousseau and Diderot,* Albany: State University of New York, and R. Wokler, ed., 1995, Rousseau and Liberty, New York: Manchester University Press.

22. Montesquieu's concept of physical nature is still tied to the mechanistic representations of his time. As part of the general introduction to his work, he writes: "Since we observe that the world, though formed by the motion of matter, and void of understanding, subsists through so long a succession of ages, its motions must certainly be directed by invariable laws; and could we imagine another world, it must also have constant rules, or it would inevitably perish" (Spirit of Laws I, 1.1).

Montesquieu qualifies his concept of nature when it comes to the conduct of human beings: "But the intelligent world is far from being so well governed as the physical. For though the former has also its laws, which of their own nature are invariable, it does not conform to them so exactly as the physical world. This is because, on the one hand, particular intelligent beings are of a finite nature, and consequently liable to error; and on the other, their nature requires them to be free agents. Hence they do not steadily conform to their primitive laws; and even those of their own instituting they often infringe" (Spirit of Laws 1.2).

The separation of Montesquieu from the French Philosophes included in this chapter is supported by Arendt (1963, p. 303):

"Montesquieu's separation of power, because it is so intimately connected with the theory of checks and balances, has often been blamed with the scientific, Newtonian spirit of the time. Yet nothing could be more alien to Montesquieu than the spirit of modern science. This spirit, it is true, is present in James Harrington and his 'balance of property' as it is present in Hobbes. No doubt this terminology drawn from the sciences carried even then a great deal of plausibility—as when John Adams praises Harrington's doctrine for being 'as infallible a maxim in politics as that action and reaction are equal in mechanics'. Still, one may suspect that it was precisely Montesquieu's political, non-scientific language which contributed so much to his influence; at any rate, it was in a non-scientific and non-mechanical spirit and quite obviously under the influence of Montesquieu that Jefferson asserted that 'the government we fought for . . . should not only be founded on free principles (by which he meant the principles of limited government) but in which the powers of government should be so divided and balanced among several bodies of magistracy as that no one could transcend their legal limits, without being effectually checked and restrained by the other" (Notes on the State of Virginia) (See Chapter 5).

23. One can construct here an example of sociopolitical conceptual continuity. To maintain a high level of productivity on barren soil may involve large scale fertilization and the building of a dam for irrigation that requires a balance of collective and individual action that is adjusted to optimal common interest defined in specific terms of fertilization and irrigation that establish conceptual continuity between the participating social units.

24. Burke's emphasis on the limitations of a purely rationalistic (deductive) representation of reality vis-à-vis the complexities of human sociopolitical existence is pointed out by Hoffman and Levack (1949).

"The error of the rationalists was their attempt to travel rationally farther than reason could lead them; hence they left the world of reality for a fanciful world of mad metaphysical abstraction. They assumed an understanding of what they lacked, the patience and humility to explore; they prescribed without diagnosing, and took their legislative remedies from the cupboard of theory, forgetting that the nature of any political entity ought to be known before one can venture to say what is fit for its conservation" (p. xvii).

Even more metaphorically, Burke expresses a similar thought in his *Reflections* (pp. 152–153), with regard to human rights:

"These metaphysic rights entering into common life, like rays of light which pierce into dense medium, are by the laws of nature, refracted from their straight line. Indeed, in the gross and complicated mass of human passions and concerns, the primitive rights of man undergo such a variety of refractions and reflections that it becomes absurd to talk of them as if they continued in the simplicity of their original direction. The nature of man is intricate; the objects of society are of the greatest possible complexity and therefore no simple disposition or direction of power can be suitable either to man's nature, or to the quality of his affairs. When I hear the simplicity of contrivance aimed at and boasted of in any new political constitution, I am at no loss to decide that the artificers are grossly ignorant of their trade, or totally negligent of their duty. The simple governments are fundamentally defective, to say no worse of them. If you were to contemplate society in but one point of view, all these simple modes of policy are infinitely captivating. In effect each would answer its single end much more perfectly than the more complex is able to attain all its complex purposes. But it is better that the whole should be imperfectly and anomalously answered, than that, while some parts are provided for with great exactness, others might be totally neglected, or perhaps materially injured, by over-care for a favourite member."

**CHAPTER 4: COMPLEXITY AS THE REALITY OF THE POSTMODERN
WORLD: CONCEPTUAL CONTINUITY IN NONIDEAL SYSTEMS**

1. Systems and chaos theories are essentially advanced computations for the identification of the specific components of complex systems. However, as pointed out by Quastler (1965), one of the founders of systems theory, "If a system is well enough known to be analyzable in every respect by physical and chemical methods then a systems theory approach does not seem necessary or rewarding." The analysis of living systems has apparently advanced to this point.

2. Recent advances in the analysis of the three-dimensional fine structure of the myosin molecule and its changes during contraction are summarized in several short reports (Stull 1996).

3. General information about mechanisms of gene expression and related cell processes can be found in several textbooks of cellular and molecular biology, such as J. Darnell, H. Lodisch, and D. Baltimore, 1990, *Molecular Cell Biology,* N.Y.: W. H. Freeman and Company; and B. Alberts, D. Bray, J. Lewis, M. Raff, K. Roberts, and J. Watson, 1997, *Molecular Biology of the Cell;* Hamden, Conn.: Garland Publishing. Reviews of more specific topics are published by the Cold Spring Harbor Laboratory Press and the CRC Press, and summaries of specific items of genetic mechanisms are included in *The Encyclopedia of Cell Biology,* J. Kendrew, ed., 1994, Cambridge: Blackwell Scientific Publications. Reports on recent advances in this field have been published in many journals..

**CHAPTER 5: CONCEPTUAL CONTINUITY AND THE
HIGH-LEVEL COMPLEXITY OF DEMOCRACY**

1. A fundamental incongruity between the ideal systems of physics and the nonideal sociopolitical systems has been emphasized by Steiner (1986, p. 217). In sciences like physics or biology: "The observer and the object of observation are clearly separated, unlike the social sciences, where both observer and observed are human beings. When social scientists publish their results, those results may be read by the objects of their study. Social scientists in this way are actors in the world they observe. How to define and measure a political variable is influenced by the political values of the researcher. With a pluralism of values of course, researchers will arrive at different results."

   The difference between the systems of classical and political science has been reviewed more fully by Zolo (1992, Chap. 2). The limitations of economic theory have been pointed out by Elster as reviewed by Alan Ryan, *New York Review of Books,* October 10, 1991, p. 19.

2. The presumptive termination of this trend was heralded by Daniel Bell as "The End of Ideology" (1962). In his words:

   "The end of ideology is not—should not be—the end of utopia as well. If anything, one can begin anew the discussion of utopia only by being aware of the trap of ideology. The point is that ideologists are 'terrible simplifiers'. Ideology makes it unnecessary for people to confront individual issues on their individual merits. One simply turns to the ideological vending machine, and out comes the prepared formula. And when these beliefs are suffused by apolcalyptic fervor, ideas become weapons, and with dreadful results.

"There is now, more than ever, some need for utopia, in the sense that men need—as they always needed—some vision of their potential, some manner of fusing passion with intelligence. Yet the ladder to the City of Heaven can no longer be a 'faith ladder', but an empirical one: a utopia has to specify *where* one wants to go, *how* to get there, the costs of the enterprise, and some realization of, and justification for, the determination of *who* is to pay" (Bell 1962, p. 405).

The end-of-ideology position has been revived by Francis Fukuyama's much-discussed *The End of History and the Last Man* (1991) as modified by Leiss (1993). The "end of ideology" theme was anticipated by Karl Popper (1962). In a poignant article "The Responsibility of Intellectuals" (*New York Review of Books,* June 22, 1995, p. 36–37) President Vaclav Havel summarizes Popper's position that is relevant in the present context because it describes the process that can be regarded as development of conceptual continuity in complex sociopolitical systems.

3. For a concise definition of democracy see Hook (1989) and for a comprehensive assessment of contemporary democracy in the United States see Dahl (1972, 1982, 1989).

4. According to Dolbeare (1981) the Constitution promised to be "a plan reflecting underlying philosophical assumptions about human nature and the purposes of social order and a deeply personal allocation of immediate economic and political burdens and benefits."

5. "If we try the Constitution by its last relation to the authority by which amendments are to be made, we find it neither wholly *national* nor wholly *federal.* Were it wholly national, the supreme and ultimate authority would reside in the majority of the people of the Union; and this authority would be competent at all times, like that of a majority of every national society, to alter or to abolish its established government. Were it wholly federal, on the other hand, the concurrence of each State in the Union would be essential to every alteration that would be binding on all. The mode provided by the Constitution is not founded on either of these principles. In requiring more than a majority, and in particular in computing the proportion by states, not by citizens, it departs from the national and advances toward the federal character; in rendering the concurrence of less than the whole number of States sufficient, it loses again the federal and partakes of the national character.

"The proposed Constitution, therefore, is, in strictness, neither a national nor a federal Constitution, but a composition of both. In its foundation it is federal, not national; in the sources from which ordinary powers of the government are drawn it is partly federal and partly national; in the operation of these powers, it is national; and finally in the authoritative mode of introducing amendments, it is neither wholly federal nor wholly national" (*Federalist Papers,* No. 39; Rossiter, 1961, p. 246).

6. "The Fathers of the Constitution, children of the Age of Reason, fabricated what we may call a Newtonian scheme of government, static rather than dynamic. Not only that, but since experience had taught them that all government was to be feared they exhausted their ingenuity in devising methods of checking governmental tyranny. They manufactured, to this end, a complicated system of checks and balances—the federal system, the tripartite division of powers, the bicameral legislature, judicial review and so forth.

"Such a system if adhered to rigorously, would result very speedily in governmental paralysis" (Commager, p. 193).

7. The term "opposites" is used here to include both dichotomy and duality. Dichotomy stands for division into two, mutually exclusive entities. Duality stands for a double nature of a single entity.

8. "The history of the last century is accordingly in large measure history of a group of financial titans, whose methods were not scrutinized with too much care and who were honored in proportion as they produced the results, irrespective of the means they used. The financiers who pushed the railroads to the Pacific were always ruthless, often wasteful, and frequently corrupt; but they did build railroads, and we have them today. It has been estimated that the American investor paid for the American railway system more than three times over in the process; but in spite of this fact the net advantage was to the United States. As long as we had free land, as long as the population was growing by leaps and bounds, as long as our industrial plants were insufficient to supply our own needs, society chose to give the ambitious man free play and unlimited regard provided only that he produced the economic plant so much desired" (in Commager 1951, p. 352).

9. According to Commager (1951, p. 337): "Lester Ward was the first major scholar to attack the inadequate science, the dubious logic, and the specious rhetoric of the Spencer-Sumner school, and he remained the ablest. To the study of sociology he brought immense resources of scientific and philosophical learning and a firm grasp of the meaning of evolution to social development."

One of the basic tenets of Ward's sociology, taken from his main work *Psychic Factors of Civilization* (1893), is: "All human institutions—religion, government, law, marriage, custom—together with innumerable other modes of regulating social, industrial, and commercial life, are broadly viewed, only so many ways of meeting and checkmating the principle of competition as it manifests itself in society. And finally, the ethical code and the moral law of enlightened man are nothing else than the means adopted by reason, intelligence, and refined sensibility for suppressing and crushing out the animal nature of man—for chaining the competitive egoism that all men have inherited from their animal ancestors" (in Commager 1951, p. 339).

The laissez-faire philosophy was critically evaluated by the American Economic Association as summarized by one of its members, Richard T. Ely:

"We hold that the doctrine of *laissez-faire* is unsafe in politics and unsound in morals, and that it suggests an inadequate explanation of the relations between the state and the citizens. In other words we believe in the existence of a system of social ethics; we do not believe that any man lives for himself alone, nor yet do we believe social classes are devoid of mutual obligations corresponding to their infinitely varied inter-relations. All have duties as well as rights and, as Emerson said several years ago it is time we heard more about duties and less about rights.

"It is asked: what is meant by *laissez-faire?* It is difficult to define *laissez-faire* categorically, because it is so absurd that its defenders can never be induced to say precisely what they mean yet it stands for a well-known, though rather vague set of ideas, to which appeal is made every day in the year by the bench, the bar, the newspapers, and our legislative bodies. It means that government, the state, the people in their collective capacity, ought not to interfere in industrial life; that, on the contrary, free contract should regulate all the economic relations of life and public authority should simply enforce this, punish crime and preserve peace. It means that the laws of economic life are natural laws like those of physics, and chemistry, and that this life must be left to the free play of natural forces. One adherent uses these words: 'This industrial world is governed by natural laws. These laws are superior to man. Respect this providential order—let alone the work of God'" (in Commager, 1951, p. 336–337).

10. The basis for this reinterpretation was provided by Woodrow Wilson, in *The State: Elements of Historical and Practical Politics* (Boston: D. C. Heath & Co., 1895); Herbert Croly, *The Promise of American Life,* Ed. Arthur M. Schlesinger, Jr. (Cambridge, Mass.: Harvard University Press, 1965; orig. ed. 1909). John Dewey, *Liberalism and Social Action* (New York: Perigee Books, 1980; orig. ed. 1935, all cited in Kessler 1993, p. 235).

11. Entitlements and other "uncontrollable" spending surpassed discretionary federal spending in 1975 and have surpassed it ever since. The figures are collected in *Economic and Budget Outlook: Fiscal Years 1993–1997.* Washington, D.C. Congressional Budget Office, 1992, p. 118); John Marini, *The Politics of Budget Control: Congress, the Presidency, and the Growth of the Administrative State,* Washington, D.C., Crane Russak, 1992. Citations from Kessler, 1993, p. 241).

12. It should be considered that during this time foreign affairs (Cold War; opening of China; the state of the South American Republics) took the center stage of administrative attention.

13. "The Perot Phenomenon illustrates the degree to which parties, and discussions of the constitutional dimension of American politics, have fallen into disrepute" (Schramm and Wilson 1993, p. vii).

"It ought to be honestly asked whether the much vaunted American Constitution deep-frozen in the late-eighteenth century when 'checks and balances' were a more important consideration than national efficiency, does not hinder—or nowadays even paralyze—the taking of unpopular but necessary reforms" (Kennedy 1990, p. 40).

"But the deeper reason why the Constitution's significance for our party system has declined is that the Constitution itself has been under running intellectual and political attack for about a hundred years. As a result the Constitution no longer provides a clear raison d'être for our parties. Deprived of their deepest ground of legitimacy, however, the parties have not prospered under the new dispensation; their public respectability has ebbed. Ross Perot's abortive presidential candidacy in 1992 was a sign of this, an ominous conjunction of disdain for political parties and barely suppressed impatience with the Constitution" (Kessler 1993, p. 231).

14. "Following Hegel, holists often cast individualism in historical terms as a falling off from some original or possible unity" (Rosenblum, 1993, pp. 77–104). "The people is not the organic body of a common and rich life" but is reduced to "an atomistic, life-impoverished multitude" (Stephen Smith *Hegel's Critique of Liberalism.* Chicago University Press, Chicago, 1989, cited in Rosenblum 1993).

15. A religious basis of holistic individualism has been adopted by Charles Taylor in *Sources of the Self: The Making of the Modern Identity* as reviewed by Bernard Williams in the *New York Review of Books,* November 8, 1990, and by Alastair MacIntyre in *After Virtue* and *Whose Justice? Which Rationality?* reviewed by Martha Nussbaum in the *New York Review of Books,* December 7, 1989.

16. See the editorial by M. B. Zuckerman, "Beware the Adversary Culture" in *U.S. News and World Report,* June 12, 1995).

17. The seven departments are Foreign Affairs, Interior, Justice and Police, Military, Finance, Public Economy, and Transport-Communication-Energy.

CHAPTER 6: THE THREAT TO SOCIOPOLITICAL COMPLEXITY: THE SIMPLIFYING SOCIOPOLITICAL IDEOLOGIES OF NATIONALISM AND COMMUNISM

1. "Revolutions or at least such major sociopolitical upheavals as the French Revolution belong to the class of historical phenomena whose significance is not to be judged by the intentions or expectations of those who make them, or even those that could be imputed to them by subsequent analysis. Such intentions are not, of course, irrelevant to the study of the phenomenon. However, they cannot determine it, because uncontrollability of process and outcome is its essential characteristic.

   "For we are not dealing with phenomena to which the criteria of social problem-solving apply more than peripherally; where human agencies can effectively choose between correct and incorrect solutions, alternative strategies, or more or less wasteful or elegant methods of achieving ends specifiable in advance. Such ends are not absent, but they are dwarfed by what is uncontrolled and unintended. Even if we suppose that the Constitution of 1791 was exactly what the leaders of the National Assembly of 1789 had intended to achieve and that it represented what turned out to be the lasting achievement of the revolution, it cannot be seriously supposed that at the time of its promulgation it was in anyone's power to declare that the revolution is over. The subsequent events proved the contrary" (Hobsbawm 1990b, pp. 31–32; see also Arendt 1963; Skocpol 1979; Skocpol and Kestenbaum 1990).

   "Simultaneously meritocratic and nationalist, the enlightened reader could no longer make his peace with the historicist particularism of the Old Regime, but he was not, for all that, prepared to construct a pluralistic policy. The breakdown of revolutionary Jacobinism is implied by the divided, schizophrenic nature of antecedent French and Enlightenment thought, which was unable to sort out the rights of the individual from the responsibilities of the citizen" (Higonnet 1990, pp. 100–101).

2. "The laws so solemnly passed between August 4 and August 11 were so much part of the Enlightenment that they were almost a natural culmination. Enjoying a profound and well prepared consensus around liberty, equality before the law, and property, subsequent civil legislation of the revolutionary assemblies and the Consulate would remain within the legal framework defined during these 'historic' days; the abolition of nobility, equality in inheritance, and limitations on the right to make a will were logical consequences of the spirit of August 4. The differences between the Convention's debates and those of the Consulate concerning the future Civil Code, finally completed in 1804, were more than minor, but the principles common to both periods were established in August 1789. On that date, in a sense, the Revolution was complete, but it had yet to be inscribed in the sovereignty of the people. In that sense it had just begun. The *establishment* of principles was an extension of the Enlightenment. The *implementation* of the principles anticipated the 19th century" (Furet 1989, p. 113).

"The ideas of the revolutionaries were the ideas of the *Philosophes* with a difference; they were the ideas of the *Philosophes* deprived of their qualifying clauses, placed in an emotional instead of an intellectual setting, and transmuted by a one-sided reading of Rousseau into something that would have deeply horrified Voltaire and that did actually alienate the few men of philosophic tradition living in 1789. . . .

"The interpretation of Rousseau as endorsing the revolution is a misinterpretation. . . .

"Given equality the revolutionaries had no desire to limit the power of government; rather did they, as Burke saw, reduce all citizens to a mass of politically undifferentiated individuals in order to concentrate authority in its hands, splitting the community artificially into the individual wills and then fusing them in an authoritarian General Will. The individualist democracy of Locke passed with the *Philosophes* into egalitarianism, and under the influence of Rousseau the latter was for a time confused with democracy, particularly in the minds of foreign observers. But the despotic principle of the French state soon reasserted itself, and the dictatorship of the people came to mean first the rule of its small oligarchy and then the tyranny of an emperor" (Cobban 1960, p. 134–136).

The unrest among the lower classes began before the *Declaration* and continued thereafter, independently of the utopian ideas. In this sense the unqualified utopias cannot be said to have caused the revolutionary movement. However, the quest for constitutional change in response to these developing social inadequacies were reinforced by utopian ideas, in particular those of Rousseau, beginning with Sieyès's *What Is the Third Estate* to Robespierre's address of 1793.

3. From the Constitution of 1791, Article 3: "The source of all sovereignty resides essentially in the nation; no group, no individual may exercise authority not emanating expressly therefrom," and Article 6: "Law is the expression of the general will; all citizens have the right to concur personally or through their representatives in its formation—it must be the same for all, whether it protects or punishes."

It is plausible that these articles became prescriptions for oppression during later phases of the Revolution, as elaborated in the following comments:

"The whole tradition of the absolute monarchy in France, in both theory and practice, was dominated by the principle that there is a single supreme power, that of the king, to which all other powers are subordinated. It was the absolute monarchy that imposed on France the idea—and not only the idea, but also the reality—that sovereign power resides in *one* place and one place only. Hence, it was not the idea of indivisible sovereignty that Rousseau bequeathed to the Revolution. Rather, it was the idea that the people is *one*, that it is possessed, like an individual, of a *single* will" (Mannin 1989, p. 839).

The *Declaration of the Rights of Man and the Citizen* is not a document of purely American inspiration. There was another powerful influence which, if still not purely French, was at least Francophone. This was the influence of Rousseau. And the influence of Rousseau worked in such a way as to nullify the Jeffersonian guarantees contained in the declaration.

Sieyès's contribution—a simple and far-reaching one—was to incorporate Rousseau's abstract and emotionally inert "general will" into the emotionally powerful concept of the nation.

Sieyès described the nation in the same terminology, and in the same style as Rousseau had described the "general will." "The nation exists before all, it is the origin of everything. Its will is always legal, it is the law itself."

"Limitations of power were out and 'full sovereignty' was in. Under the influence of Sieyès and Rousseau, the Rights of Man were turning into the Rights of Leviathan" (O'Brien 1990, p. 48).

"Once we note that the declaration of the *Rights of Man and the Citizen* carries with it the absolutist and lethal principle of the general will, then the declaration cannot be the benign and liberatory document so splendidly celebrated in Paris in July 1989" (O'Brien 1990, I:90).

The relationship of Rousseau to Sieyès and to the revolutionary precepts requires further consideration in view of Mannin's contribution cited above.

4. Yet, had France remained at peace with the rest of Europe, it is possible that, in spite of such disruption, the Revolution might have stopped its course or, at least, not been carried far beyond the settlement of 1791 (Rude 1988, p. 73).

5. "Herder upholds the value of variety and spontaneity, of the different, idiosyncratic paths pursued by peoples, each with its own style, ways of feeling and expression, and denounces the measuring of everything by the same timeless standards—in effect, those of the dominant French culture, which pretends that its values are valid for all time, universal, immutable. One culture is no mere step to another. Greece is not an antechamber to Rome. . . . If each culture expresses its own vision and is entitled to do so, and if goals and values of different societies and ways of life are not commensurable, then it follows that there is no single set of principles, no universal truth for all men and times and places" (Berlin 1998, pp. 567–568).

"The establishment of one world, organised on universally accepted rational principles—the ideal society—is not acceptable. Kant's defence of moral freedom and Herder's plea for the uniqueness of cultures, for all the former's insistence on rational principles and the latter's belief that national differences need not lead to collisions, shook—some might say undermined—what I have called the three pillars of the main Western tradition" (Berlin 1998, p. 568). ("To all genuine questions there is one true answer and one only . . . the true answers to such questions are in principle knowable . . . these true answers cannot clash with one another . . . together these answers must form a harmonious whole" [Berlin 1998, p. 555].)

6. "Nationalism, unlike tribal feeling or xenophobia, to which it is related, but with which it is not identical, seems scarcely to have existed in ancient or classical times. . . . Its emergence as a coherent doctrine may perhaps be placed and detected in the last third of the eighteenth century in Germany, more particularly in the conceptions of the Volksgeist and the Nationalgeist in the writings of the vastly influential poet and philosopher Johann Gottfried Herder (Berlin 1991, pp. 243–244). Herder's thought is dominated by his conviction that among the basic needs of men, as elemental as that for food or procreation or communication is the need to belong to a group. More fervently and imaginatively than Burke and with the wealth of historical and psychological examples, he argued that

every human community had its own unique shape and pattern. Its members were born in a stream of tradition which shaped their emotional and physical development no less than their ideas. There was a central historically developing pattern that characterized the life and activity of every identifiable community and, most deeply, that unit which by his own time had come to be the nation (Berlin 1991, p. 244). Herder and his disciples believed in the peaceful co-existence of a rich multiplicity and variety of national forms, the more diverse the better, but it turned into embittered, aggressive nationalist self-assertion" (Berlin 1998, p. 568).

7. "Not of the reign of feeling, but of the assertion of the will—the will to do what is universally right in Kant, but something which cuts even deeper in the case of Herder: the will to live one's own regional, local life, to develop one's own unique (*eigentümlich*) values" (Berlin 1998, p. 568).

8. "Fichte is the true father of romanticism, above all in his celebration of will over calm, discursive thought . . . self-awareness springs from encountering resistance" (Berlin 1998, p. 569).

"Fichte goes further; values, principles, moral and political goals, are not objectively given, not imposed on the agent by nature or a transcendent God; I am not determined by ends: the ends are determined by me. Food does not create hunger, it is my hunger that makes it food" (Berlin 1998, pp. 569–570).

"Fichte's will is dynamic reason, reason in action. Yet it was not reason that seems to have impressed itself upon the imagination of his listeners in the lecture-halls of Jena and Berlin, but dynamism, self-assertion; the sacred vocation of man is to transform himself and his world by his indomitable will. . . . ends are not . . . objective values. . . . Ends are not discovered at all, but made, not found but created" (Berlin 1998, p. 571).

"Man is not a mere compounder of pre-existent elements; imagination is not memory, it literally generates, as God generated the world. There are no objective rules, only what we make" (Berlin 1998, p. 571).

"From here it is no great distance to the worlds of Byron's gloomy heroes—satanic outcasts, proud, indomitable, sinister—Manfred, Beppo, Conrad, Lara, Cain—who defy society and suffer and destroy. They may, by the standards of the world, be accounted criminal, enemies of mankind, damned souls: but they are free; they have not bowed the knee in the House of Rimmon; they have preserved their integrity at a vast cost in agony and hatred. The Byronism that swept Europe, like the cult of Goethe's *Werther* half a century earlier, was a form of protest against real or imaginary suffocation in a mean, venal and hypocritical milieu, given over to greed, corruption and stupidity. Authenticity is all: 'the great object in life', Byron once said, 'is Sensation'" (Berlin 1998, p. 572).

9. In Fichte's conception the individuum has a dual character. The final aim of individual existence is self-realization. However, this involves an adjustment of the freedom of the individual self to the freedoms of others, an awareness of relationships that can be subsumed under the notion of love. Only by voluntarily limiting his own freedom with respect to that of others does the individual attain his own full freedom. A community is formed of ethically free individuals. In contrast, the freedom of human beings in their natural state is adjusted by a system of laws that is required to maintain man's social status (see Hahn 1969, pp. 93–100).

CHAPTER 7: CONCLUSION

1. Whatever generalizations pertaining to human affairs existed in the past half century, they did not predict the most decisive world events: the development and use of the atomic bomb, the Cold War between former allies, the stepwise resolution of the Cold War atomic bomb threat, the dissolution of the Soviet Union; the resurgence of virulent nationalism; the increase in crime, violence, and drug addiction in the United States; the military and diplomatic debacle of the Vietnam War; the economic ascendancy of the Asiatic nations; the initiation of the most decisive political moves (use of atom bomb, ending of the Second World War, Marshall Plan, United Nations support for Korean War, Truman Doctrine for the containment of Soviet ambitions) by a president (Truman) who started as a midwestern farmhand and haberdasher; the Ross Perot candidacy for presidency; the merger of the computer rivals, IBM and Apple.

2. The importance of specific factors and conditions that control major sociopolitical developments is indicated by the uniqueness of interacting individuals and the identification of sometimes peripheral economic or geopolitical conditions, which create an initial area of common interest—the equivalent of conceptual continuity—between initially unrelated sociopolitical entities. Pertinent examples are seen in the following events:

   The threat of an atomic holocaust was averted by several successive interactions, beginning with the Kennedy-Khrushchev test ban, Gorbachev-Reagan missile number reduction, Gorbachev-Bush maximal dismantling; continuing negotiations toward unification of Europe; the Mandela-Clark agreement abandoning South Africa's apartheid principle. Establishment of peace between Israel and Egypt through negotiations mediated by President Carter at Camp David and initiation of a stepwise rapprochement between Israel and the Palestinians after years of secret negotiations by nonofficial individuals with the prospect of a resolution of half a century of conflict; the most-favored nation negotiations between the United States and China; the protracted efforts toward settlement of the Balkan conflict; the efforts toward establishment of the health care and welfare system in the United States; the differences in the conflict resolution processes in different localities (Abu-Nimer 1996).

3. The leading organization promoting discussion groups is the Topsfield Foundation's "Study Circles Resource Center" (P.O. Box 203, Pomfret, CT, 06258, Tel. 860–928–2616). The Center provides advice on how to organize and conduct discussions together with literature references to specific topics and a syllabus summarizing the material.

References

Abrahams, A. M., and S. L. Shapiro. 1990. The Wave Makers. Modeling the Relativistic Storms that Roil Through Space-Time. *The Sciences* 30:30–36.

Abu-Nimer, M. 1996. Conflict Resolution Approaches: Western and Middle Eastern Lessons and Possibilities. *American Journal of Economics and Sociology* 55(1):35–51.

Acton, H. B. 1972. Hegel, Georg Wilhelm Friedrich. In: *Encyclopedia of Philosophy.* P. Edwards, ed., New York: Macmillan, vol. 3, pp. 435–451.

Aida, S. et al. 1985. *The Science and Praxis of Complexity.* Tokyo: United Nations University.

Anderson, F. H. 1948. *The Philosophy of Francis Bacon.* Chicago: University of Chicago Press.

Arendt, H. 1963. *On Revolution.* New York: Viking Press.

Arndt, E. M. 1908. *Ausgewählte Werke.* Vols. 1–16. H. Meisner and R. Geerds, eds., vols. 9–12 Geist der Zeit., vol. 12 Verfassung und Pressfreiheit, pp. 40–102. Das Turnwesen, pp. 180–195. Leipzig: Max Hess Verlag.

Attali, M. J. 1990. Lives on the Horizon: A New Order in the Making. *New Perspectives Quarterly,* Spring, p. 6.

Baker, K. M. 1972. Condorcet, Marquis De. In: *The Encyclopedia of Philosophy.* P. Edwards, ed., New York: Macmillan, vol. 2, pp. 182–184.

———. 1975. Condorcet. *From Natural Philosophy to Social Mathematics.* Chicago: University of Chicago Press.

————. 1989. Condorcet. In: *Critical Dictionary of the French Revolution.* F. Furet and M. Ozouf, eds. Cambridge, Mass.: Belknap Press of Harvard University Press.

————. 1991. *Inventing the French Revolution: Essays of French Political Culture in the Eighteenth Century.* Cambridge: Cambridge University Press.

Barber, B. 1984. *Strong Democracy: Participatory Politics in a New Age.* Berkeley: University of California Press.

Barker, E. 1925. *Greek Political Theory.* London: Methuen.

Barnard, F. M. 1988. *Self-Direction and Political Legitimacy: Rousseau and Herder.* Oxford: Clarendon Press.

Barnes, H. E. 1941, rpt. 1965. *An Intellectual and Cultural History of the Western World.* New York: Dover Publications, Inc.

Becker, C. L. 1932. *The Heavenly City of the Eighteenth Century Philosophers.* New Haven: Yale University Press.

Bell, D. 1962. *The End of Ideology.* New York: Free Press.

Bellah, R. N., R. Madsen, W. M. Sullivan, A. Swindler, and S. M. Tipton. 1985. *Habits of the Heart.* Berkeley: University of California Press.

————. 1991. *The Good Society.* New York: Alfred A. Knopf.

Berlin, I. 1956. *The Age of Enlightenment. The 18th Century Philosophers.* A Mentor Book, New York: New American Library, Inc.

————. 1991. *The Crooked Timber of Humanity: Chapters in the History of Ideas.* New York: Alfred A. Knopf.

————. 1996. On Political Judgment. *New York Review of Books.* October 3, pp. 26–30.

————. 1998. *The Proper Study of Mankind: An Anthology of Essays.* H. Hardy and R. Hausheer. New York: Farrar, Straus and Giroux.

Bertalanffy, L. von. 1960. Principles and Theory of Growth. In: *Fundamental Aspects of Normal and Malignant Growth.* W. W. Nowinski, ed., Amsterdam: Elsevier Publishing Company.

Bethke-Elshtain, J. 1995. *Democracy on Trial.* New York: Basic Books (HarperCollins).

Bjerkness, J. 1969. Atmospheric Teleconnections from the Equatorial Pacific. *Monthly Weather Rev.* 97:163–172.

Blanpied, W. A. 1969. *Physics: Its Structure and Evolution.* Waltham: Blaisdell Publishing Company.

Bobbio, N. 1989. *Democracy and Dictatorship. The Nature and Limits of State Power.* Minneapolis: University of Minnesota Press.

Bohman, J. 1996. *Public Deliberation: Pluralism, Complexity and Democracy.* Cambridge, Mass.: MIT Press.

Bondanella, J.C. 1988. Rousseau: A Biographical Sketch. In: *Rousseau's Political Writings.* A. Ritter and J. C. Bondanella, eds., New York: W. W. Norton.

Brinton, C. 1972. Enlightenment. In: *The Encyclopedia of Philosophy,* New York: Macmillan and Free Press, vol. 2., pp. 519–525.

Brooks, R. C. 1927. *Government and Politics of Switzerland.* New York: World Book Company.

Burke, E. 1790. Reflections on the Revolution in France and on the Proceedings in Certain Societies in London Relative to That Event. In: *The Works of the Right Honourable*

Edmund Burke. A New Edition, vol. 5, pp. 27–438. London: F. and C. Rivington, 1815. The same work edited and with Introduction by Conor Cruise O'Brien, 1969, Hermondsworth and Middlesex: Penguin Books.

Butt, P. A. 1980. European Nationalism in the Nineteenth and Twentieth Centuries. In: *The Roots of Nationalism: Studies in Northern Europe.* R. Mitchison, ed. Edinburgh: John Donald Publishers, Ltd., pp. 1–10.

Butterfield, H. 1965. *The Origins of Modern Science.* New York: Macmillan.

Butzer, K. W. 1980. Civilization: Organization or System? *American Scientist* 68:517–523.

Cairns, J., G. S. Stent, and J. D. Watson, eds. 1966. *Phage and the Origins of Molecular Biology.* Cold Spring Harbor, New York: Cold Spring Harbor Laboratory.

Calode, C. 1988. *Theories of Computational Complexity.* Amsterdam, North Holland: Imprint Elsevier Science.

Campbell, J. 1982. *Grammatical Man: Information, Entropy, Language, and Life.* New York: Simon and Schuster.

Capers, G. M. 1989. United States. 26. Sectional Conflict and Preservation of the Union, 1815–1877. In: *Encyclopedia Americana,* vol. 27, pp. 726–744.

Carens, J. H., ed. 1993a. *Democracy and Possessive Individualism: The Intellectual Legacy of C. B. Macpherson.* Albany: State University of New York Press.

———. 1993b. Possessive Individualism and Democratic Theory: Macpherson's Legacy. In: Carens, J. H., ed. *Democracy and Possessive Individualism. The Intellectual Legacy of C.B. Macpherson.* Albany: State University of New York Press.

Cartwright, N. 1989. *How the Laws of Physics Lie.* Oxford: Oxford University Press.

Cassirer, E. 1955. *The Philosophy of the Enlightenment. Second Edition.* Translated by F. C. A. Koellin and J. P. Pettegrove. Princeton, N.J.: Princeton University Press.

Casti, J. 1979. *Connectivity, Complexity and Catastrophe in Large Scale Systems.* New York: John Wiley and Sons.

———. 1994. *Complexification.* New York: HarperCollins.

Chancellor, J. 1990. *Peril and Promise: A Commentary on America.* New York: Harper and Row.

Chartier, R. 1991. *The Cultural Origins of the French Revolution.* Durham, N.C.: Duke University Press.

Clapp, J. G. 1972. John Locke. In: *The Encyclopedia of Philosophy.* P. Edwards, ed., New York: Macmillan and Free Press, vol. 4, pp. 487–503.

Cobban, A. 1934. *Rousseau and the Modern State.* London: George Allen and Unwin.

———. 1960. *Edmund Burke and the Revolt Against the Eighteenth Century, Second Edition.* London: George Allen and Unwin; New York: Barnes and Noble.

Codding, Jr., G. A. 1983. The Swiss Political System and the Management of Diversity. In: *Switzerland at the Polls. The National Elections of 1979.* H. R. Penniman, ed. Washington: American Enterprise Institute for Public Policy Research, pp. 1–29.

Cohen, I. B. 1976. The Eighteenth Century Origins of the Concept of Scientific Revolution. *History of Ideas,* vol. 37, p. 257.

———. 1980. *The Newtonian Revolution, With Illustrations of the Transformation of Scientific Ideas.* Cambridge: Cambridge University Press.

———. 1985. *The Birth of a New Physics.* New York: W. W. Norton.

Cohler, A. M. 1988. *Montesquieu's Comparative Politics and the Spirit of American Constitutionalism.* Lawrence: University Press of Kansas.

Commager, H. S., ed. 1951. *Living Ideas in America.* New York: Harper and Brothers.

Cook, W. J., and J. Schrof. 1990. *U.S. News and World Report* 108 (12):52–58.

Craig, G. A. 1992. Above the Abyss. Review of A. Bullock. Hitler and Stalin: Parallel Lives. *New York Review,* April 9, 1992, p. 3.

Crozier, M., S. P. Huntington, and J. Watanuki. 1975. *The Crisis of Democracy.* New York: New York University Press.

Dahl, R. A. 1972. *Democracy in the United States: Promise and Performance.* New Haven: Yale University Press.

———. 1982. *Dilemmas of Pluralist Democracy: Autonomy versus Control.* New Haven: Yale University Press.

———. 1989. *Democracy and Its Critics.* New Haven: Yale University Press.

Davis, W. N. 1989. United States. 27. The Age of Industrial Growth, 1877–1919. In: *Encyclopedia Americana,* vol. 27, pp. 745–745r.

Delbrück, M. 1949. A Physicist Looks at Biology. *Trans. Conn. Acad. Sci. 38*:173–190.

Demos, R. 1937. *Introduction to the Dialogues of Plato.* Translated by B. Jowett. New York: Random House.

Denton, T. A. et al. 1990. Fascinating Rhythm: A Primer of Chaos Theory and Its Application to Cardiology. *Am. Heart J.* 120(6, Pt. 1):1419–1440.

Deutsch, K. W. 1966. Recent Trends in Research Methods in Political Science. In: *A Design for Political Science: Scope, Objectives, and Methods.* J. C. Charlesworth, ed. Philadelphia: The American Academy of Political and Social Science, Monograph #6.

———. 1969. *Nationalism and Its Alternatives.* New York: Alfred A. Knopf.

Dewey, J. 1927. *The Public and Its Problems.* New York: Henry Holt and Company.

———. 1929. *The Quest for Certainty: A Study of the Relation of Knowledge and Action.* New York: Minton, Balch.

Diamandopoulos, P. 1972. Anaximenes. In: *The Encyclopedia of Philosophy.* P. Edwards, ed., New York: Macmillan and Free Press, vol. 1, pp. 118–119.

Diem, A. 1993. Switzerland. In: *Encyclopedia Brittanica* 28:341–356.

Dietz, M. G., ed. 1990. *Thomas Hobbes and Political Theory.* Lawrence: University of Kansas Press.

Dietz, W. 1980. *Johann Gottfried Herder. Abriss seines Lebens und Schaffens.* Berlin: Aufbau-Verlag.

Dolbeare, K. M. 1981. *American Political Thought.* Monterey, Calif.: Duxbury Press Wadsworth, Inc.

Dressler, A. 1989. In the Grip of the Great Attractor. *The Sciences,* September/October, pp. 28–34.

Drucker, P. F. 1969. *The Age of Discontinuity.* New York: Harper and Row Publishers.

Durant, W. 1939. *The Life of Greece.* New York: Simon and Schuster.

Durkheim, E. 1960. *Montesquieu and Rousseau; Forerunners of Sociology.* Ann Arbor: University of Michigan Press.

Edwards, P., ed. 1972. *The Encyclopedia of Philosophy.* New York: Macmillan and Free Press.

Eigen, M., and R. Winkler. 1981. *The Laws of the Game: How the Principles of Nature Govern Change*. New York: Alfred A. Knopf.

Eisenberg, A. I. 1995. *Reconstructing Political Pluralism*. Albany: State University of New York Press.

Eisenberg, E., and T. L. Hill. 1985. Muscle Contraction and Free Energy Transduction in Biological Systems. *Science* 227:999–1006.

Emanuel, K. A. 1988. Toward a General Theory of Hurricanes. *American Scientist* 77:371–379.

Engels, F. 1940. *Dialectics of Nature*. London: Lawrence and Wishart.

Fawcett, D. W. 1986. *Bloom and Fawcett: A Textbook of Histology*, 11th Edition. Philadelphia: W. B. Saunders.

Featherstone, J. 1978. Rousseau and Modernity. In: *Daedalus, Proceedings of the American Academy of Arts and Sciences* 107/3, pp. 167–192.

Feher, F. 1990. Introduction. In: *The French Revolution and the Birth of Modernity*. F. Feher, ed. Berkeley: University of California Press, pp. 1–10.

Feynman, R. P. 1985, rpt. 1988. *QED. The Strange Theory of Light and Matter*. Princeton, N.J.: Princeton University Press.

Fichte, J. G. 1846. *Sämmtliche Werke*. J. H. Fichte, ed. Berlin: Veit und Comp.

————. 1846. Reden an die Deutsche Nation. In: *Sämmtliche Werke*. J. H. Fichte, ed., vol. 7, pp. 259–499. Berlin: Veit und Comp.

Flood, R. I., and E. R. Carson. 1988. *Dealing with Complexity*. New York: Plenum Press.

Freedman, D. H. 1990. Makers of Worlds. *Discovery*, July, pp. 46–52.

Fukuyama, F. 1992. *The End of History and the Last Man*. New York: Free Press.

Furet, F. 1968. *Interpreting the French Revolution*. Translated by Elborg Forster. Cambridge: Cambridge University Press.

Furet, F., and Mona Ozouf, eds. 1989. *A Critical Dictionary of the French Revolution*. Cambridge, Mass.: Belknap Press of Harvard University Press.

Furley, D. J. 1972. Parmenides of Elea. In: *The Encyclopedia of Philosophy*. P. Edwards, ed. New York: Macmillan and Free Press, vol. 6, pp. 47–51.

Gabbey, A. 1990. Newton and Natural Philosophy. In: *Companion to the History of Modern Science*. R. C. Olby, G. N. Cantor, J. R. R . Christie and M. J. S. Hodge, eds., London: Routledge, pp. 243–263.

Galbraith, J. K. 1977. *The Age of Uncertainty*. Boston: Houghton Mifflin.

Gay, P. 1969. *The Enlightenment: An Interpretation*. 2 vols. New York: Alfred A. Knopf.

Gleick, J. 1987. *Chaos. Making a New Science*. New York: Viking.

Glotz, P. 1990. *Der Irrweg des Nationalstaats*. Stuttgart: Deutsche Verlags-Anstalt.

Goldman, R. M. 1989. The Democratic Party. In: *Encyclopedia Americana*, vol. 7, p. 693–698.

Gottinger, H. W. 1983. *Coping with Complexity*. Dordrecht and Boston: D. Reidel.

Graubard, S. R., ed. 1978. *Rousseau for Our Time*. Daedalus. Proceedings of the American Academy of Arts and Sciences, vol. 107/3.

Grimsley, R. 1972. Rousseau, Jean-Jacques. In: *The Encyclopedia of Philosophy*. New York: Macmillan and Free Press, vol. 7, pp. 218–225.

Grüne, E., and K. J. Pitterle. 1983. Switzerland's Political Parties. In: *Switzerland at the Polls.*

H. R. Penniman, ed., Washington, D.C.: The National Elections of 1974. The American Enterprise Institute for Public Policy Research, pp. 30–54.

Guthrie, W. K. C. 1981. *A History of Greek Philosophy.* Vol. 6 of *Aristotle an Encounter.* Cambridge: Cambridge University Press, chap. 1.

Gutmann, A., and D. Thompson. 1996. *Democracy and Disagreement: Why Moral Conflict Cannot be Avoided in Politics and What Should Be Done About It.* Cambridge, Mass.: Belknap Press of Harvard University Press.

Habermas, J. 1996. *Between Facts and Norms.* Cambridge, Mass.: MIT Press.

Hahn, K. 1969. *Staat, Erziehung und Wissenschaft.* München: Verlag Beck.

Haldane, J. S. 1931. *The Philosophical Basis of Biology.* Donnelan Lectures in the University of Dublin 1930. London: Hodden and Stoughton.

Hamby, A. L. 1989. United States. 29. The Modern Nation: Progress and Travail, 1945– . In: *Encyclopedia Americana* vol. 27, pp. 748–748i.

Harold, F. M. 1986. *The Vital Force: A Study of Bioenergenetics.* New York: W. H. Freeman and Company.

Harre, R. 1986. *The Physical Sciences since Antiquity.* New York: St. Martin's Press, chap.1.

Harrington, F. H. 1989. United States. 25. The Founding of the Nation, 1763–1819. In: *Encyclopedia Americana* vol. 27, pp. 708–725.

Herbert, G. B. 1989. *Thomas Hobbes. The Unity of Scientific and Moral Wisdom.* Vancouver: University of British Columbia Press.

Herrmann, H. 1989. The Unification of Muscle Structure and Function: A Semicentennial Anniversary. *Perspectives in Biology and Medicine* 33, No. 1, pp. 1–11.

Herzstein, R. E. 1974. *Adolf Hitler and the German Trauma 1913–1945. An Interpretation of the Nazi Phenomenon.* New York: Putnam.

Higonnet, P. 1990. Cultural Upheaval and Class Formation During the French Revolution. In: *The French Revolution and the Birth of Modernity.* F. Feher, ed., Berkeley: University of California Press, pp. 69–102.

Hill, A. V. 1950. A Challenge to Biochemistry. In: *Metabolism and Function.* D. Nachmansohn, ed., New York: Elsevier Publishing Company, Inc.

Hobbes, T. 1973. *Leviathan.* Introduction by K. R. Minogue. Everyman's Library Series. New York: Dutton.

Hobsbawm, E. 1990a. *Nations and Nationalism since 1870.* Cambridge: Cambridge University Press.

———. 1990b. The Making of a "Bourgeois Revolution." In: *The French Revolution and the Birth of Modernity.* F. Feher, ed., Berkeley: University of California Press, pp. 30–48.

Hoffman, R. J. S., and P. Levack. 1949. Burke's Politics. In: *Selected Writings and Speeches of Edmund Burke on Reform, Revolution, and War.* New York: Alfred A. Knopf. (The quotation to "Works" refers to *The Works of the Right Honorable Edmund Burke* 1865–1867, Boston: Little, Brown and Company.)

Hook, S. 1975. *Revolution, Reform and Social Justice.* New York: New York University Press.

———. 1983. *Marxism and Beyond.* Totowa, N.J.: Rowman and Littlefield.

———. 1989. Democracy. In: *Encyclopedia Americana* vol. 7, pp. 684–691.

Hopkins, F. G. 1949. The Dynamic Side of Biochemistry. In: *Hopkins and Biochemistry.* J. Needham and E. Baldwin, eds., Cambridge: W. Heffer and Sons, Ltd.

Hroch, M. 1968. *Vorkämpfer der nationalen Bewegung bei den kleinen Völkern Europas.* Prague: Karlova University.

Hussey, E. 1971. *The Presocratics,* chap. 1. New York: Charles Scribner's Sons.

Ignatieff, M. 1984. *The Needs of Strangers.* New York: Penguin Books.

Jahn, F. L. 1991. *Deutsches Volkstum* (German Nationhood). Berlin: Aufbau-Verlag.

Jowett, B., transl. 1920. *The Dialogues of Plato.* Oxford: Oxford University Press.

Kahn, C. H. Anaximander. 1972. In: *The Encyclopedia of Philosophy.* P. Edwards, ed., New York: Macmillan and Free Press, vol. 1, pp. 117–118.

———. 1972. Empedocles. In: *The Encyclopedia of Philosophy.* P. Edwards, ed., New York: Macmillan and Free Press, vol. 2, pp. 496–499.

Kaplan, R. D. 1994. The Coming Anarchy. *Atlantic Monthly,* 273/2, February, pp. 44–76.

Kauffman, S. A. 1993. *The Origin of Order. Self Organization and Selection in Evolution.* New York: Oxford University Press.

Keane, J. 1989. Introduction: Democracy and the Decline of the Left. In: *Democracy and Dictatorship.* N. Bobbio. Minneapolis: University of Minnesota Press.

———. 1993. Stretching the Limits of the Democratic Imagination. In: *Democracy and Possessive Individualism. The Intellectual Legacy of C.B. Macpherson.* J. H. Carens, ed. Albany: State University of New York Press, pp. 105–135.

Kekes, J. 1993. *The Morality of Pluralism.* Princeton, N.J.: Princeton University Press.

Kellert, S. H. 1993. *In the Wake of Chaos.* Chicago: University of Chicago Press.

Kennan, G. F. 1989. The History of Arnold Toynbee. *New York Review of Books,* June 1, pp. 19–21.

Kennedy, P. 1987. *The Rise and Fall of Great Powers.* New York: Random House.

———. 1990. Fin-de-Siècle. *New York Review of Books,* June 28, pp. 31–40.

———. 1993. *Preparing for the Twenty-first Century.* New York: Random House.

Kerferd, G. B. 1972. Aristotle. In: *Encyclopedia of Philosophy,* P. Edwards, ed., New York: Macmillan and Free Press, vol. 1, pp. 151–162.

———. 1972. Anaxagoras of Clazomenae. In: *Encyclopedia of Philosophy,* P. Edwards, ed., New York: Macmillan and Free Press, vol. 1, pp. 115–117.

Kessler, C. R. 1993. Political Parties, The Constitution and the Future of American Politics. In: *American Political Parties and Constitutional Politics.* P. W. Schramm and B. P. Wilson, eds., Lanham, Md.: Rowman and Littlefield Publishers, Inc.

Kirk, G. S., and J. E. Raven. 1960. (2nd ed., 1984). *The Presocratic Philosophers.* Cambridge: Cambridge University Press.

Kitcher, P. 1988. Explanation and Scientific Understanding. In: *Theories of Explanation.* J. C. Pitt, ed., pp. 167–187, New York: Oxford University Press.

Klir, G. J. 1972. Preview. The Polyphonic General Systems Theory. In: *Trends in General Systems Theory.* G. J. Klir, ed., New York: Wiley Interscience.

Koepke, W. 1987. *Johann Gottfried Herder* (Engl.). New York: Simon and Schuster.

Kohn, H. 1956. *Nationalism and Liberty. The Swiss Example.* London: George Allen and Unwin.

———. 1961. *The Idea of Nationalism. A Study in Its Origin and Background.* New York: Macmillan.

———. 1965. *Nationalism: Its Meaning and History.* Princeton, N.J.: D. Van Nostrand Company, Inc. Anvil Books.

————. 1967. *Prelude to Nation-States. The French and German Experience, 1789–1815.* Princeton, New York: D. Van Norstrand Company, Inc.

Kraynak, R. P. 1990. *History and Modernity in the Thought of Thomas Hobbes.* Ithaca, N.Y.: Cornell University Press.

Kuhn, T. S. 1957. *The Copernican Revolution: Planetary Astronomy in the Development of Western Thought.* Cambridge, Mass.: Harvard University Press.

Ladd, E. C. 1993a. Change and Continuity in American Values in the 90s. Institute for Public Relations Research and Education.

————. 1993b. The Advance of Democratic Politics. London: *Times Literary Supplement.* September 29.

————. 1993–1994. The Myth of Moral Decline. *Rights and Responsibilities* 4(1):52–68.

————. 1994. The American Ideology. An Exploration of the Origins, Meaning, and the Role of American Values. *Occasional Papers and Monograph Series* #1. Storrs, Conn.: Roper Center for Public Opinion Research.

Ladd, Jr., E. C. 1970. *American Political Parties. Social Change and Political Response.* New York: W. W. Norton and Company, Inc.

Ladd, Jr., E. C., and C. D. Hadley. 1975. *Transformations of the American Party System. Political Coalitions from the New Deal to the 1970s.* New York: W. W. Norton & Company, Inc.

Leigh, R. A. 1988. Liberty and Authority on Social Contract. In: *Rousseau's Political Writings.* A. Ritter and J. C. Bondanella, eds., New York: W. W. Norton, pp. 232–243.

Leiss, W. 1993. The End of History and Its Beginning Again; or, The Not-Quite-Yet-Human Stage of Human History. In: *Democracy and Possessive Individualism. The Intellectual Legacy of C.B. Macpherson.* J. H. Carens, ed., New York: State University of New York Press, pp. 263–274.

Loeb, J. 1912. *The Mechanistic Conception of Life.* Chicago: University of Chicago Press.

————. 1916. *The Organism as a Whole.* New York: G. P. Putnam's Sons.

Lukes, S. 1973a. *Individualism.* Oxford: Basil Blackwell.

————. 1973b. Types of Individualism. In: *Dictionary of the History of Ideas.* P. P. Weiner, ed., vol. II, pp. 594–604. New York: Charles Scribner's Sons.

MacIntyre, A. 1984. *After Virtue: A Study in Moral Theory.* Notre Dame, Ind.: University of Notre Dame Press.

————. 1989. *Whose Justice? Which Rationality?* Notre Dame, Ind.: University of Notre Dame Press.

Mannin, B. 1989. Rousseau. In: *A Critical Dictionary of the French Revolution.* F. Furet and M. Ozouf, eds., Cambridge, Mass.: Belknap Press of Harvard University Press, pp. 829–841.

Mansbridge, J. J. 1980. *Beyond Adversarial Democracy.* New York: Basic Books.

————., ed. 1990. *Beyond Self-Interest.* Chicago: University of Chicago Press.

————. 1990. The Rise and Fall of Self-Interest in the Explanation of Political Life. In: *Beyond Self-Interest.* J. J. Mansbridge, ed., Chicago: University of Chicago Press, pp. 3–24.

Mansfield, Jr., H. C. 1993. Political Parties and American Constitutionalism. In: *American*

Political Parties and Constitutional Politics. P. W. Schramm and B. P. Wilson, eds., Lanham, Md.: Rowman and Littlefield Publishers, Inc.

Manuel, F. E. 1978. A Dream of Eupsychia. In: *Rousseau for Our Time.* Daedalus, Proceedings of the American Academy of Arts and Sciences, 107/3, pp. 1–12.

Margenau, H. 1950. *The Nature of Physical Reality.* New York: McGraw-Hill.

Marx, K. 1970. *Contribution to the Critique of Political Economy.* M. Dobb, ed., New York: International Publishers.

McKeon, R., ed. 1941. *The Basic Works of Aristotle.* New York: Random House.

Medawar, P. B. 1967. *The Art of the Soluble.* London: Methuen and Company, Ltd., pp. 21–38.

Menand, L. 1992. The Real John Dewey. *New York Review of Books,* June 25, pp. 50–55.

Merry, H. J. 1970. *Montesquieu's System of Natural Government.* West Lafayette, Ind.: Purdue University Studies.

Meyer, A. 1934. *Ideen und Ideale der Biologischen Erkenntnis.* Barth: Leipzig.

Mitchison, R., ed. 1980. *The Roots of Nationalism: Studies in Northern Europe.* Edinburgh: John Donald Publishers, Ltd.

Molnar, P. 1989. The Geologic Evolution of the Tibetan Plateau. *American Scientist* 77:350–360.

Monastersky, R. 1990a. Forecasting into Chaos. *Science News* 137:280–282.

———. 1990b. Tibet's Tectonic Escape Act. Caught in a Closing Vice between India and Asia, Tibet May Slip Out to the Side. *Science News* 138:24–26.

Montesquieu, B. de 1949. *The Spirit of the Laws.* Translated by T. Nugent. Introduction by F. Newmann, ed. New York: Hafner Publishing Company.

Morgan, N. 1990. From Physiology to Biochemistry. In: *Companion to the History of Modern Science.* R. C. Olby, G. N. Cantor, J. R. R. Christie, and M. J. S. Hodge, eds. London: Routledge, pp. 494–502.

Morin, E. 1985. On the Definition of Complexity. In: *The Science and Praxis of Complexity.* Tokyo: United Nations University, pp. 62–68.

Morowitz, H. J. 1968. rpt. 1979. *Energy Flow in Biology.* Woodbridge, Conn.: OxBow Press.

Morris, R. 1987. The Nature of Reality. The Universe after Einstein. Farrar, Straus and Giroux, New York: Noonday Press.

Murchland, B. 1971. *The Age of Alienation.* New York: Random House.

Nance, R. D., T. R. Worsley, and J. B. Moody. 1988. The Supercontinent Cycle. *Scientific American* 259:72–79.

Nau, H. R. 1992. *The Myth of America's Decline: Leading the World Economy into the 1990s.* New York: Oxford University Press.

Nettles, C. P. 1989. United States. 24. European Exploration and Settlement. *Encyclopedia Americana* 27:692–707.

Neumann, F., ed. 1949. *The Spirit of the Laws by Baron de Montesquieu.* New York: Hafner Publishing Company.

Nisbet, H. B. 1979. *Herder and the Philosophy and History of Science.* Cambridge: The Modern Humanities Research Association.

Nisbet, R. 1988. Rousseau and Equality. In: *Rousseau's Political Writings.* A. Ritter and J. C. Bondanella, eds., New York: W. W. Norton, pp. 244–259.

Nora, P. 1989. Nation. In: *A Critical Dictionary of the French Revolution.* F. Furet and M. Ozouf, eds., Cambridge, Mass.: Belknap Press of Harvard University Press, pp. 742–753.

North, J. 1995. Why Western Europe: The Danger of Seeking a Single Explanation for the Rise of Science. London: *Times Literary Supplement,* December 15, p. 3.

Nye, Jr., J. S. 1991. *Bound to Lead: The Changing Nature of American Power.* New York: Basic Books.

Ober, J. 1989. *Mass and Elite in Democratic Athens.* Princeton, N.J.: Princeton University Press.

O'Brien, C. C. 1990. The Decline and Fall of the French Revolution. A Review of a Critical Dictionary of the French Revolution. F. Furet and M. Ozouf, eds., *New York Review of Books,* February 15, pp. 46–51.

―――. 1992. *The Great Melody: A Thematic Biography and Commented Anthology of Edmund Burke.* Chicago: University of Chicago Press.

Ogle, W. 1912. De Partibus Animalium. In: *The Works of Aristotle.* Smith, J. H. and W. D. Ross, eds., vol. 5. Oxford: Clarendon Press.

Olby, R. C. 1990. The Molecular Revolution in Biology. In: *Companion to the History of Modern Science.* R. C. Olby, G. N. Cantor, J. R. R. Christie and M. J. S. Hodge, eds. London: Routledge, pp. 503–520.

Orear, J. 1967. *Fundamentals of Physics.* New York: John Wiley and Sons, Inc.

Padover, S. K., ed. 1971. On Revolution. In: *The Karl Marx Library* vol. 1. New York: McGraw-Hill.

Papentin, F. 1980. On Order and Complexity. 1. General Considerations. *J. Theor. Biol.* 87:421–456.

Penniman, H. R. 1983. *Switzerland at the Polls. The National Elections of 1979.* Washington: American Enterprise Institute for Public Policy Research.

Peratt, A. L. 1990. Not with a Bang. The Universe May Have Evolved from a Vast Sea of Plasma. *The Sciences,* January/February, pp. 24–32.

Peters, R. S. 1972. Hobbes. In: *The Encyclopedia of Philosophy.* P. Edwards, ed., New York: Macmillan and Free Press, vol. 4, pp. 30–46.

Philander, G. 1989. El Niño and La Niña. *American Scientist* 77:451–459.

Pitt, J. C., ed. 1988. *Theories of Explanation.* New York: Oxford University Press.

Popper, K. 1962. *The Open Society and Its Enemies.* New York: Harper.

Porter, T. M. 1990. Natural Science and Social Theory. In: *Companion to the History of Modern Science.* R. C. Olby, G. N. Cantor, J. R. R. Christie and M. J. S. Hodge, eds. London: Routledge.

Pundt, A. G. 1935. *Arndt and the Nationalist Awakening in Germany.* New York: Columbia University Press.

Quastler, H. 1965. General Principles of System Analysis. In: *Theoretical and Mathematical Biology.* T. H. Waterman and H. J. Morowitz, eds., New York: Blaisdell Publishing Company, p. 332.

Ramos, V. A. 1989. The Birth of South America. *American Scientist* 77:444–451.

Randall, Jr., J. H. 1960. *Aristotle,* chap. 1. New York: Columbia University Press.

Ravetz, J. R. 1990. The Copernican Revolution. In: *Companion to the History of Modern*

Science. R. C. Olby, G. N. Cantor, J. R. R. Christie, and M. J. S. Hodge, eds. London: Routledge, pp. 201–216.

Rawls, J. 1971. *A Theory of Justice.* Cambridge, Mass.: Belknap Press of Harvard University Press.

Richter, M. 1977. *The Political Theory of Montesquieu.* Cambridge: Cambridge University Press.

Riesman, D. 1961. *The Lonely Crowd. A Study of the Changing American Character.* New Haven, Conn.: Yale University Press.

Ritter, A, and J. C. Bondanella, eds. 1988. *Rousseau's Political Writings.* Translated by J. C. Bondanella. New York: W. W. Norton and Company.

Roohan, J. E. 1989. United States. 28. Economic Boom, Depression and War 1919–1945. In: *Encyclopedia Americana* vol. 27, pp. 746–746m.

Rosenberg, A. 1985. *The Structure of Biological Science.* Cambridge: Cambridge University Press.

Rosenblum, N. L. 1989. *Liberalism and the Moral Life.* Cambridge, Mass.: Harvard University Press.

―――. 1993. The Individualist/Holist Debate and Bentham's Claim to Sociological and Psychological Realism. In: *Democracy and Possessive Individualism. The Intellectual Legacy of C. B. Macpherson.* J. H. Carens, ed. New York: State University of New York Press, pp. 77–104.

Rossiter, C., ed. 1961. *The Federalist Papers.* New York: New American Library (Penguin Books).

Rude, G. 1988. *The French Revolution.* New York: Weidenfeld and Nicolson.

Ryan, A. 1992. Who Was Edmund Burke? *New York Review of Books,* December 3, pp. 37–42.

Ryle, G. 1972. Plato. In: *Encyclopedia of Philosophy.* P. Edwards, ed. New York: Macmillan and Free Press, vol. 6, pp. 314–333.

Sagan, E. 1991. *The Honey and the Hemlock: Democracy and Paranoia in Ancient Athens and Modern America.* New York: Basic Books.

Samuel, R., ed. 1989. *Patriotism: The Making and Unmaking of British National Identity.* London: Routledge, Chapman and Hall.

Sandel, M. J. 1996. *Democracy's Discontent: America in Search of a Public Philosophy.* Cambridge, Mass.: Belknap Press of Harvard University Press.

Sattler, R. 1986. *Biophilosophy: Analytical and Holistic Perspectives.* Berlin: Springer Verlag.

Sauder, G., ed. 1987. *Johann Gottfried Herder. 1744–1803* (German). Hamburg: Felix Meiner Verlag.

Schaffer, S. 1990. Newtonianism. In: *Companion to the History of Modern Science.* R. C. Olby, G. N. Cantor, J. R. R. Christie, and M. J. S. Hodge, eds., London: Routledge, pp. 610–626.

Schenkel, E. 1933. *Individualität und Gemeinschaft. Der Demokratische Gedanke bei J.G. Fichte.* Zurich: Rascher & Cie. A.-G., Verlag, Leipzig, and Stuttgart.

Schorske, C. E. 1979. *Fin-De-Siècle-Vienna.* New York: Alfred A. Knopf.

Schramm, P. W., and B. P. Wilson, eds. 1993. *American Political Parties and Constitutional Politics.* Lanham, Md.: Rowman and Littlefield Publishers, Inc.

Schrödinger, E. 1944. rpt. 1962. *What Is Life: The Physical Aspect of the Living Cell.* Cambridge: Cambridge University Press.

Schumpeter, J. 1987. *Capitalism, Socialism, and Democracy.* London: Allen and Unwin.

Silk, J., A. S. Szalay, and Y. B. Zeldovich. 1983. The Large-Scale Structure of the Universe. *Scientific American* 249:72–80.

Simon, H. A. 1962. The Architecture of Complexity. *Proc. Am. Philos. Soc.* 106:467–482.

Skocpol, T. 1979. *States and Social Revolutions: A Comparative Analysis of France, Russia, and China.* Cambridge: Cambridge University Press.

Skocpol, T., and M. Kestenbaum. 1990. Mars Unshackled: The French Revolution in World Historical Perspective. In: *The French Revolution and the Birth of Modernity.* F. Feher, ed., Berkeley: University of California Press, pp. 13–29.

Smith, C. 1990. Energy. In: *Companion to the History of Modern Science.* R. C. Olby, G. N. Cantor, J. R. R., Christie and M. J. S. Hodge, eds., London: Routledge, pp. 326–341.

Soedjatmoko. 1985. Opening Statement. In: *The Science and Praxis of Complexity.* Tokyo: United Nations University, pp. 1–6.

Soros, G. 1997. The Capitalist Threat. *Atlantic Monthly* 279(2):45–58.

Starobinski, J. 1988. The Political Thought of Jean-Jacque Rousseau. In: *Rousseau's Political Writings.* A. Ritter and J. C. Bondanella, eds., New York: W. W. Norton, pp. 221–231.

Steiner, J. 1983. Conclusion: Reflections on the Consociational Theme. In: *Switzerland at the Polls. The National Elections of 1979.* H. R. Penniman, ed. Washington: American Enterprise Institute for Public Policy Research, pp. 161–177.

———. 1986. *European Democracies.* New York: Longman, Inc.

Stokes, M. C. 1972. Heraclitus of Ephesus. In: *The Encyclopedia of Philosophy.* P. Edwards, ed., New York: Macmillan and Free Press, vol. 3, pp. 477–481.

Strauss, L. 1952/1963, 1973. *The Political Philosophy of Hobbes.* Chicago: University of Chicago Press.

Strohman, R. C. 1997. Epigenesis and Complexity: The Coming Kuhnian Revolution in Biology. *Nature Biotechnology* 15(3):194–200.

Stull, J. T. 1996. Myosin Minireview Series. *J. Biol. Chem.* 271(27):15849.

Sussman, G. J., and J. Wisdom. 1988. Numerical Evidence That the Motion of Pluto is Chaotic. *Science* 241:433–437.

———. 1992. Chaotic Evolution of the Solar System. *Science* 257:56–62.

Tainter, J. A. 1988. *The Collapse of Complex Societies.* Cambridge: Cambridge University Press.

Talmon, J. L. 1952. *The Origins of Totalitarian Democracy.* London: Secker & Warburg.

Tamny, M. 1990. Atomism and Mechanical Philosophy. In: *Companion to the History of Modern Science.* R. C. Olby, G. N. Cantor, J. R. R., Christie and M. J. S. Hodge, eds. London: Routledge, pp. 597–609.

Taylor, C. 1989. *Sources of the Self: The Making of Modern Identity.* Cambridge, Mass.: Harvard University Press.

Thompson, D. W. 1916/1942. *Growth and Form.* New York: Cambridge University Press and Macmillan.

Toulmin, S. 1990. *Cosmopolis: The Hidden Age of Modernity.* New York: Free Press.

Touma, J., and J. Wisdom. 1993. The Chaotic Obliquity of Mars. *Science* 259:1294–1297.

Tsanoff, R. A. 1972. Johann Gottlieb Fichte. In: *Encyclopedia of Philosophy*. P. Edwards, ed., New York: Macmillan, vol. 3, pp. 192–196.

Vartanian, Aram. 1972. Claude-Adrien Helvetius. In: *The Encyclopedia of Philosophy*. P. Edwards, ed., New York: Macmillan and Free Press, vol. 3, pp. 471–473.

———. 1972. Paul-Henri Thiry, Baron D'Holbach. In: *The Encyclopedia of Philosophy*. P. Edwards, ed. New York: Macmillan and Free Press, vol. 4, pp. 49–51.

———. 1972. Julien Offray de La Mettrie. In: *The Encyclopedia of Philosophy*. P. Edwards, ed., New York: Macmillan and Free Press, vol. 4, pp. 379–382.

Voss, J. F., J. A. Lawrence, and R. A. Engle. 1991. From Representation to Decision. An Analysis of Problem Solving in International Relations. In: *Complex Problem Solving: Principles and Mechanisms*. R. J. Sternberg and P. A. French, eds., Hillsdale, N.J.: Lawrence Erlbaum Associates.

Walker, G. T., and E. W. Bliss. 1932. World Weather. *V. Mem. Royal Meteorol. Soc.* 4: 53–84.

Watkins, J. W. N. 1965. *Hobbes' System of Ideas. A Study in the Political Significance of Philosophical Theories.* London: Hutchinson University Library.

Weaver, W. 1948. Science and Complexity. *American Scientist* 36:536–544.

Weinberg, S. 1977. *The First Three Minutes.* New York: Basic Books.

———. 1993. *Dreams of a Final Theory.* New York: Pantheon Books.

Westbrook, R. B. 1992. *John Dewey and American Democracy.* Ithaca, N.Y.: Cornell University Press.

White, S. K., ed. 1995. *The Cambridge Companion to Habermas.* Cambridge: Cambridge University Press.

Whitehead, A. N. 1925 (1941). *Science and the Modern World.* New York: Macmillan.

Zolo, D. 1992. *Democracy and Complexity.* University Park: Pennsylvania State University Press.

Permissions

Grateful acknowledgment is made to the following publishers for permission to reprint excerpts from the listed references.

Addison-Wesley Educational Publishers Inc.: *European Democracies* by J. Steiner. Copyright © 1986.

American Association for the Advancement of Science: "Muscle Contraction and Free Energy Transduction in Biological Systems" by E. Eisenberg and T. L. Hill, *Science* 227, 999–1006, Fig. 1, 1985.

Cambridge University Press: *The Collapse of Complex Societies* by J. Tainter. Copyright © 1988. *The Political Theory of Montesquieu* by M. Richter. Copyright © 1977.

Chapman and Hall: *Bloom and Fawcett: A Textbook of Histology* by D. W. Fawcett, Fig. 10-18, 1986 (11th edition; 12th edition 1994 available).

Columbia University Press: *Arndt and the Nationalist Awakening in Germany* by A. G. Pundt. Copyright © 1935. *Aristotle* by J. H. Randall. Copyright © 1960.

Connecticut Academy of Arts and Sciences: "A Physicist Looks at Biology" by M. Delbrück in *Transactions Connecticut Academy of Science 38,* 1949.

Daedalus, Journal of the American Academy of Arts and Sciences: Issue entitled "Rousseau for Our Time," summer 1978, Vol. 107, No. 3, including the articles "A Dream of Eupsychia" by F. E. Manuel and "Rousseau and Modernity" by J. Featherstone.

Dutton Signet: *The Federalist Papers* by Alexander Hamilton et al., introduced by C. Rossiter. Copyright © 1961 by New American Library.

Farrar, Straus & Giroux, Inc. (New York) and Curtis Brown Ltd. (London): *The Proper Study of Mankind* by I. Berlin. Copyright © 1998.

HarperCollins: *Living Ideas in America* by H. S. Commager. © 1951. *The Honey and the Hemlock: Democracy and Paranoia in Ancient Athens and Modern America* by E. Sagan. © 1991.

Harvard University Press: *End of Ideology* by D. Bell. Harvard University Press, republication edition, 1988. *A Critical Dictionary of the French Revolution* by F. Furet and M. Ozouf, eds. Copyright © 1989.

Alfred A. Knopf Inc. (New York) and John Murray (London): *The Crooked Timber of Humanity* by I. Berlin. Copyright © 1991.

Laurence & Wishart Ltd. London: *Dialectics of Nature* by F. Engels (Preface: J. B. S. Haldane), 1940.

Methuen Publishers: *Greek Political Theory* by E Barker. [1925] 1977.

The New York Review of Books: "The History of Arnold Toynbee" by G. Kennan. Copyright © 1989. "The Decline and Fall of the French Revolution: A Review of *A Critical Dictionary of the French Revolution*" by C. C. O'Brien. Copyright © 1990.

W. W. Norton & Company, Inc.: *Rousseau's Political Writings* by A. Ritter and J. C. Bondanella. © 1988.

Oxford University Press: *The Dialogues of Plato,* translated by B. Jowett (New York: Random House, 1937).

Polity Press: Introduction to *Democracy and Dictatorship* (Bobbio) by J. Keane. © 1989.

Princeton University Press: *The Strange Theory of Light and Matter* by R. P. Feynman. © 1985. *Mass and Elite in Democratic Athens* by J. Ober. © 1989.

Routledge: "Natural Science and Social Theory" by T. M. Porter in *Companion to the History of Modern Science,* R. C. Olby et al., eds. © 1990.

Rowman and Littlefield Publishers, Inc.: "Political Parties and American Constitutionalism" by H. C. Mansfield, Jr. and "Political Parties, the Constitution, and the Future of American Politics" by C. R. Kessler, both in *American Political Parties and Constitutional Politics,* by P. W. Schramm and B. P. Wilson, eds. © 1993.

Simon and Schuster: *The Life of Greece* by W. Durant. 1939.

University of British Columbia Press: *Thomas Hobbes: The Unity of Scientific and Moral Wisdom* by G. B. Herbert © 1989.

University of California Press: *The French Revolution and the Birth of Modernity,* edited by F. Feher. Copyright © 1990.

University of Chicago Press: *Condorcet: From Natural Philosophy to Social Mathematics* by K. M. Baker. © 1975.

University of Michigan Press: *Montesquieu and Rousseau: Forerunners of Sociology* by E. Durkheim. © 1960.

Viking Penguin: *On Revolution* by H. Arendt. Copyright © 1963 by H. Arendt. *The Needs of Strangers* by M. Ignatieff. Copyright © 1985 by M. Ignatieff.

Wadsworth Publishing Company: *American Political Thought* by K. M. Dolbeare. © 1981.

Index

DATE DUE

GAYLORD			PRINTED IN U.S.A.